The American Institute of Architects

Energy Planning for Buildings

Library of Congress Catalog Card
Number 79-50796

ISBN 0-913962-08-2

AIA Catalog Number 4-M720

Manufactured in the United States
of America.

The American Institute of Architects

Energy Planning for Buildings

Michael M. Sizemore, AIA
Henry O. Clark, AIA
William S. Ostrander, PE

Preface

What is narrowly perceived as one person's crisis may be broadly considered as another's opportunity. To reduce energy use in buildings requires us to broaden our vision and consider the consequences of our otherwise isolated decisions.

This book provides a method with which to view a building, its occupants and activities as elements in a pattern of energy flow. Rather than treating energy analysis and design in a piecemeal manner, the authors have looked at the interconnected influences of climate, occupants, systems, equipment and operation procedures. By uncovering these relationships, we have sought to make possible a basic understanding of what they mean for improving energy use in new and existing buildings.

Thus each design choice has to be made with an awareness of how it affects the other parts of a building. ***That is the essence of energy-efficient design as well as great architecture.***

We hope to lead fellow architects and engineers more deeply into an energy-efficient practice. We hope further to provide to those already so involved a useful method of energy planning which they may not know about. Therefore the book may be read and understood without reviewing the calculations, or the reader can use it in all its detail, culminating in the sample problem.

The approach challenges the energy planner to view a building as a dynamic entity consisting of interrelated parts and participants. These may be used to manipulate and alter the natural energy flows that act upon the building. Translating this conceptual view of energy flow into architecture brings our built environment into harmony with the natural environment and its cycles.

The Role of The American Institute of Architects

In 1974, the AIA began a wide-ranging program designed to capture the full energy-saving potential of buildings. Two major goals were a demonstration of how energy could be saved in existing buildings and the creation of information and tools for design professionals.

Steps to reach the first goal got under way in 1976 with an in-depth analysis of the energy performance of the AIA's own five-year-old headquarters building in Washington, DC. Sizemore & Associates, architects and energy planners, was chosen to carry out this task. As a result of the study, the AIA has underwritten and completed physical modifications and operational procedures to the building to reduce its energy use by nearly 50 percent.

At about that time, the AIA decided to produce a document that would allow design professionals to help their clients save energy in buildings, with emphasis on those already built. In 1977, the AIA joined with the authors in the development and publication of this book, which Sizemore & Associates had already begun to write for its own use.

The AIA thereupon appointed a review team, and the final form of this work owes much to the time and effort of those dedicated individuals, consisting of Huber H. Buehrer, AIA, P. Richard Rittelmann, AIA and Douglas Stenhouse, AIA, AIP, and to the suggestions and comments of Herbert Epstein, FAIA, David Perkins, FAIA, Charles Ince, David C. Bullen, AIA and Joseph A. Demkin, AIA, who served as the AIA staff coordinator for the project.

Also involved in publication of this work were the following, who not only performed well on a tight schedule, but were a pleasure to work with: Stephen A. Kliment, FAIA, editorial consultant; Jeffry Corbin, graphic design and printing management; Lorna L. Palmer, manuscript coordination and preparation; Christine W. Sizemore, PhD, staff editor, and Bradford Y. Fletcher, PhD, bibliographic consultant.

Sections of this book were reviewed by George Hightower, P.E., Paul Kennon, FAIA, Ramon Noya, P.E., Dick Depta, AIA, Bob Michaud, P.E., and Herb Millkey, AIA.

Finally, we appreciate the encouragement and support of CRS, architects, planners, engineers, with whom Sizemore & Associates merged in mid-1978.

Michael M. Sizemore, AIA
Atlanta, January 1979

Overview

This book is designed to show architects, engineers and owners how best to evaluate and improve a building's energy performance. The book focuses on existing buildings. The methodology applies equally to the design of new buildings.

The book deals with every aspect of an energy planning project, from preparing the original proposal to monitoring the results. The book also describes a manual energy estimating technique and a sample problem illustrating the calculations. The manual technique will help readers evaluate any design solution for energy implications as well as provide the context for judging the value of solar-assisted alternatives. It further provides a basis for understanding current computer-based estimating techniques.

The book is written in a way that follows the chronological steps of a typical energy planning project. By providing such a framework, the authors offer a needed perspective even though the reader's own project may not follow the steps exactly.

The process for improving a building's energy performance has three logical phases: 1) study present building performance, 2) identify opportunities for improvement, and 3) evaluate these opportunities.

Executive summaries at the beginning of each chapter provide a quick feeling for what is involved in an energy planning project and help the reader decide which topics to pursue in detail. A reference list is included at the back of the book should the reader like to extend the study.

Items that may be seen as "non-energy related," such as user comfort, environmental impact, or visual appearance, may be as important as energy performance. The energy planner must identify these at the outset and study them throughout the project.

The result will be a building which is not only energy efficient, but also performs better in those other areas.

By and large the book focuses on those aspects of energy planning which are discussed in only a piecemeal manner in works published to date.

The authors have concentrated on providing a methodology and sequence of steps for professionals who 1) understand thermal load calculations, but who are just now beginning to look into energy planning, and 2) those who would like to compare their own methods with those discussed here. Other readers will be able to follow most of the discussion and gain an overall appreciation of energy planning. Specific references to other sources are few because much of the information is new and drawn from the authors' practical experience.

Following is a brief look at each chapter:

Chapter 1, **Introductory concepts,** brings up basic ideas which are new, usually misunderstood, or which the energy planner needs to be reminded of before performing the analysis.

Chapter 2, **The team: roles and responsibilities,** reviews contributions of the various "team" members on an energy planning project. Included are the architect or engineer acting as energy planner, the owner, building users, operators, energy suppliers, product manufacturers and building officials. There is also a brief look at the main points to consider when preparing the Owner/Energy Planner contract.

Chapter 3, **Study of present building performance,** the energy audit and energy analysis, shows how to study a building's present performance so the impact of any proposed change may be evaluated. Presented is a short method, the energy audit, which can be done with few calculations, and a detailed method, the energy analysis, which provides a more complete and accurate picture of a building's energy use. Various types of energy estimates are considered, and the "modified bin method" for manual estimates is discussed in detail.

Chapter 4, **Identifying opportunities,** describes the most efficient way of identifying practical opportunities for improving a building's existing performance. The chapter shows how to use the energy audit and energy analysis to best advantage.

Chapter 5, **Evaluating opportunities,** points out how to narrow the list of opportunities to those that deserve to be carried out in the form of actual modifications. Some of these will be easy to evaluate, others will call for detailed investigation. Therefore, this chapter looks at evaluation in two ways – a "quick" evaluation and a "detailed" one.

Chapter 6, **Implementation and follow-through,** shows what is involved in carrying out the energy planner's recommendations and in monitoring and maintaining the building to obtain best performance.

In the **Sample Problem,** an example of the "modified bin method," manual energy estimating technique is applied to a typical twenty-story office building. It follows the same format as that given in Chapter 3. Required calculations are presented so they can be easily followed by anyone wanting to learn the technique. Calculation forms that would commonly be drawn up by the energy planner for a specific job are used, as they are more practical than forms claiming to be universally applicable.

For clarity, the sample problem is separated from the rest of the text to let the reader first master the typical steps of an energy planning project from beginning to end without the distraction of calculations.

Readers seriously interested in learning to apply the "modified bin method" are assumed to have a working knowledge of one of the available HVAC design manuals. Other readers may only care about a general appreciation of the technique; they will need to study the sample problem less thoroughly. Nor will they need to follow every calculation. Although the sample problem appears at the end, brief examples are used throughout.

In the interest of fairness, sexist references have been kept to a minimum. In cases where the term "he" has been used to avoid contorted syntax, the word "she" may be freely substituted and vice versa.

An appendix contains detailed information relevant to the main text. There is a glossary of common terms and a selection of references for further study.

Contents

Chapter 1: Introductory Concepts

1 Linkages
Comfort
2 Space use and programming
3 Building envelope
8 Ventilation
9 Lighting
12 Energy Sources
14 Heating, ventilating, and air conditioning
15 Building thermal loads
16 Part load conditions
Load and system capacity balance
17 Equipment performance
21 Energy estimating methods
Equivalent full-load hour method
Degree-day method
22 Hour-by-hour method
Outside temperature bin method
Computer programs
22 Summary

Chapter 2: The Team: Roles and Responsibilities

24 Introduction
The energy planner
The owner
26 The building's users
The building's operators
Energy suppliers, contractors and product manufacturers
Building officials
27 The owner/energy planner agreement
Determining the owner's expectations and resources
Developing an approach to the project

Chapter 3: Study of Present Building Performance: the Energy Audit and the Energy Analysis

30 Introduction
Performing the energy audit
31 Collecting data for the energy audit
33 Studying energy performance
39 Studying energy consumption
41 Performing the energy analysis
Developing the analysis procedure
Collecting data for the energy analysis
44 Estimating energy performance
63 Preparing additional profiles
Estimating energy costs

Chapter 4: Finding Opportunities for Improved Performance

66 Introduction
Listing opportunities already identified
The energy audit and energy analysis as sources
Product manufacturers as sources
67 Publications and checklists as sources
67 Looking for additional opportunities
Identifying the owner's primary concerns
68 Identifying the largest sources of energy consumption
69 Identifying the largest sources of energy demand
71 Examining all significant variables
Reducing the job to be done
Reducing time spent doing the job
72 Increasing efficiency with which the job is done
Reducing cost per unit of energy used to do the job

Chapter 5: Evaluating Opportunities

74 Introduction
Evaluation procedure
75 Performing a quick evaluation
Establishing evaluation criteria
76 Predicting the impact of each modification
79 Evaluating each opportunity using established criteria
Listing and ranking opportunities
79 Performing the detailed evaluation

Chapter 6: Implementation and Followthrough

82 Introduction
Implementation
Organizing the team
83 Preparing contract documents
Preparing user's and operator's manuals
84 Followthrough
Checking performance
Maintenance
Monitoring

86 **Sample Problem**

Appendix

132 A System response
134 B Cost-benefit study

136 **Glossary**

142 **References**

Introductory Concepts

This chapter introduces those concepts basic to energy use in buildings which the energy planner needs to keep in mind before performing an analysis.

The various sections focus on comfort requirements as well as on the natural and artificial means of meeting them. Included are concern for comfort; space use and programming; the building envelope; ventilation; lighting; energy sources; and heating, ventilating, and air-conditioning (HVAC).

This material is an important resource when analyzing a building's energy needs in the *energy audit* and *energy analysis* described in later chapters. A final section deals with the chief estimating methods used in computing a building's expected energy needs.

Linkages

Buildings are designed and built to meet the needs of people and their activities. These needs include thermal comfort, visual comfort, privacy and communication. A building can help to make work easier and, by creating comfort conditions, invigorate its occupants.

Resources to meet the activity needs include the building itself, the energy used, and skills at hand to operate the building. A shortage in one area may be made up by an abundance in another. Thus, a lack of energy can be compensated by more skillful operation or an improved building.

For example, occupants will adjust their lighting in favor of daylighting, whereas building designers tend to assume that occupants will balk at this. The building accordingly is provided with needlessly high levels of energy-intensive lighting at every possible location.

Unfortunately, a shortage of energy may also lead to substandard conditions for occupants. For instance, an open office furnishing system, installed to reduce energy use for air distribution and electric lighting, may reduce acoustical and visual privacy. **Efficient energy use should not – and need not – be achieved at the expense of comfort.** On the other hand, involving occupants in adjusting their physical surroundings creates a healthy sense of possession of those surroundings: they enjoy being able to adjust lights, alter temperatures and open windows.

As designers consider the use of energy, they become more considerate of the needs of the occupants, resulting in improved comfort. In the past, it has been easier to overlight a space with cheap energy – on the theory that more light was somehow better than not enough – than to light it just right.

Energy use stems from energy flows between various building components which reject, absorb, produce and "filter" energy. The energy planner's challenge is to control these natural flows of energy to advantage, and to create the conditions that will allow the activity within the building to be extended to times when outdoor conditions would otherwise restrict such activity.

Efficient energy use incorporates some or all of the following approaches:

Collecting the natural energy (heat, light, nocturnal cooling, air movement) that flows through a building, and storing it for use when needed (such as storing in the floor mass the abundant solar heat that comes through a south-facing window).

Conserving collected energy (such as with double-glazing of a south-facing window).

Using waste heat from building processes, building equipment and power generation equipment.

Using renewable energy sources that function independently of the building design (such as "active" solar heating and cooling, solar energy from photovoltaics, wind, geothermal and tidal energies).

Efficiently converting non-renewable energy sources into forms required by the building (such as carefully selected furnaces and chillers).

The lines between these approaches are in practice blurred because the key to energy-efficient design is not in the type of approach but in the way they are combined. The designer must synchronize available energy with energy that is needed, using the building to control and integrate these energy flows.

How Much Heat the Body Produces

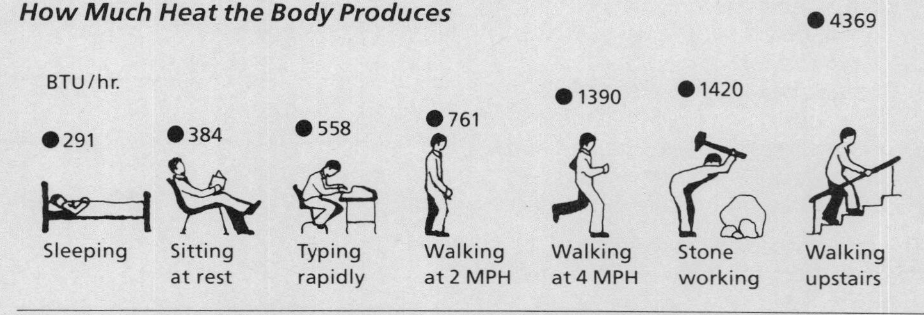

Figure 1-1 *Reprinted with the permission of Billboard Publications, Inc.*

Comfort

The human body produces heat at various rates, depending on the intensity of physical and mental activity **(figure 1-1)**.

A state of thermal equilibrium is reached whenever heat produced as a by-product of activity or effort to stay warm is dissipated at the same rate at which it is produced. Discomfort results when this state is achieved through great effort. Comfort results when the state is achieved with minimal effort.

The body's thermostat sends signals to the body to respond to environmental conditions. The response may be to produce more heat, step up blood circulation to carry off heat more rapidly for cooling, open the skin pores for sweating, or reduce activity. The ways in which the body adjusts combine to establish a balance among the four forms of body heat loss:

- **Evaporative loss** is due to sweating and breathing. At high temperatures, this method is effective but physically demanding.
- **Conducted loss** transfers heat through direct contact with physical objects, such as feet on floors.
- **Convected heat loss** occurs through contact of a surface, such as skin, with moving air.
- **Radiant heat loss** is proportional to the fourth power of the temperature differences between the body surface and surrounding objects separated from the body by space.

The totals based on these forms of heat loss from the body are illustrated in **figure 1-2**.

Forms of Heat Loss from the Human Body under Varying Temperatures

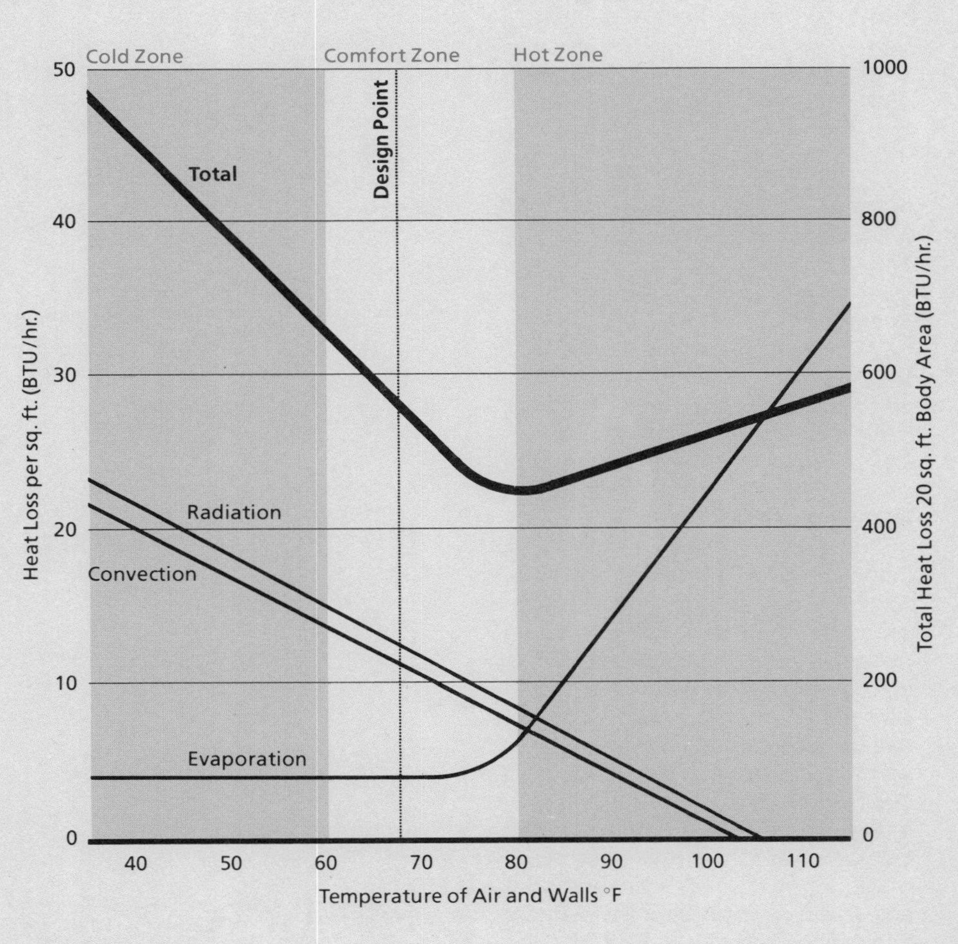

Figure 1-2 *Reprinted with the permission of Billboard Publications, Inc.*

Environmental conditions that influence body heat loss consist of:

Air movement, which increases evaporation rates and convection rates of heat loss.
Temperature of objects, which affects radiant heat losses and conducted losses.
Moisture content, which affects convection and evaporation.
Air temperature, which affects convection and evaporation.

The linkages are best seen on the comfort chart **(figure 1-3),** which shows the effect on comfort of dry bulb temperatures, humidity, air movement, and solar gain. By stressing these effects, **one may extend comfort conditions beyond the comfort envelope** into cooler temperatures so long as the occupant is surrounded by mainly warm radiant surfaces.

For example, if the occupant sits close to a large glass area, the following conditions, though different, can result in the same perceived comfort with the same outdoor winter temperature:

76°F indoors with single-glazing,
68°F with double-glazing
60°F or less temperature with a radiant heating system and double-glazing.

Comfort results from the accumulation of temperatures of surrounding surfaces and air temperature. If surfaces are made warmer, then the air can be cooler. The designer must consider the greatest comfort benefit per unit of energy and cost expended whether it be in the building shell or the heating system.

This design temperature is critical in estimating energy consumption, sizing HVAC equipment, and designing the glass conditions.

Research shows that people do not like, nor perform as well, in a constant environment as in a varied one. So the designer's task of providing a changing range of thermal and visual comfort conditions to accommodate occupants under changing external environmental conditions is in fact less restrictive than if he had to design for constant conditions, A good example is the shifting quality of lighting stemming from the use of daylight.

Changing comfort standards in a controlled manner can save a lot of energy by allowing an indoor temperature range from 68°F to 80°F in a climate with large diurnal temperature swings. One architectural firm (Burt, Hill, Kosar, Rittelmann, Associates) was able to estimate 46% savings on its "Minimum Energy Dwelling" in California compared with a constant temperature setting for that climate.

The designer's purpose is to moderate, delay, and control the impacts of climate so they are synchronized with the needs for energy established by the various activities in the building.

Space use and programming

The impact on an occupant of different and changing climatic forces will vary with the occupant's location in the building and with the building's design. This is in turn reflected in the demand for energy.

Energy-efficient space use means providing spaces which meet the needs of the building functions and which modify the impact of climate, thereby providing comfort for occupants' activities while using the least amount of off-site energy. Efficient-space use also reduces the required size of mechanical equipment.

A building's orientation has a big impact on its energy use. In summer, western exposures incur the greatest solar impact late in the afternoon when outdoor temperature is highest, whereas eastern exposures receive the most solar impact in the morning when outdoor air temperatures are cooler and hence less of a problem. Designers must therefore avoid placing on the west side of a building those spaces requiring extensive glass and large air-conditioning loads. Otherwise, the activity within that space should be moved before overheating occurs. The impacts characteristic of various exposures provide great opportunities to the alert building designer.

A key factor of energy conservation is to design spaces which provide only as much conditioned space as needed for programmed activities. Spaces used only for part of a day for one activity may often be used for others with similar comfort requirements.

The energy planner must accordingly consider trade-offs among spaces designed for a number of uses and spaces tailored for a specific type of work. As energy becomes more costly, people will be willing to accept smaller, but more carefully designed spaces. This should lead to reduced initial as well as operational HVAC system costs.

Human Comfort Zone as it Relates to Outdoor Conditions

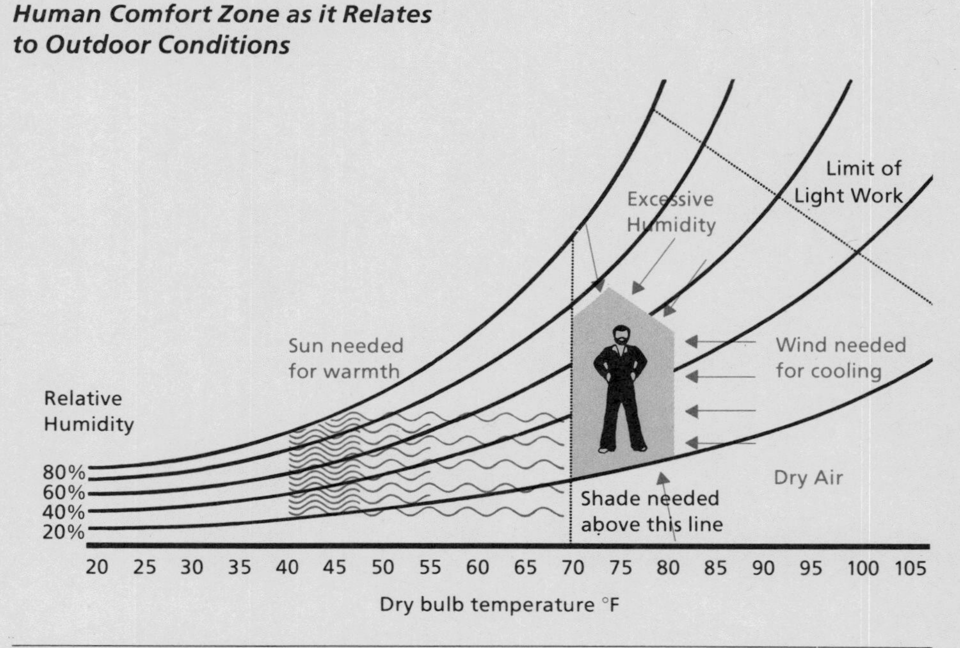

Figure 1-3 *Burt Hill Kosar Rittelmann Associates*

Specific climates result in different **building configurations,** ranging from linear buildings for warm climates (to increase cross-ventilation) to cube-like forms (to reduce heat loss) for colder climates. Spaces which require uniform year-round constant temperatures and humidity might be located underground. Those that require higher levels of light may be located on the top floor or at the periphery of a building, whereas spaces in need of infrequent, lower levels of light are placed at the core of a structure, on lower floors, or below grade.

The real goal is to integrate external natural energy flows with internal comfort needs, as illustrated in figure 1-4.

Building envelope

The envelope is the boundary between a conditioned indoor space and the outdoors. It is made up of those walls, windows, openings, roofs and floors of buildings that are exposed to the elements.

The envelope affects energy consumption in the following ways:

It reduces and impedes heat flow into or out of a building, thereby moderating the extremes of outside temperature.

It controls air movement into the building.

It reduces vapor flow into or out of a building to acceptable levels.

It controls solar heat gain.

It controls daylight entering the building.

In other words, **the envelope acts as a filter;** it does not prevent transmission of heat or vapor; it does reduce peaks and, ideally, delays and synchronizes these energy flows to bring them to acceptable, more uniform, gradually fluctuating levels to keep a building comfortable. Likewise, daylight may be filtered and directed to illuminate tasks within a building.

The building envelope can also delay peak energy use to **mesh with the local utility's power generation pattern.** The energy planner will prefer to have a building hit peak demand when other buildings do not, thereby profiting from lower off-peak rates that are becoming common and reducing the utility's need for additional expensive power plants.

There are several salient variables in the way heat flows through a building envelope.

Heat flows through the envelope as a result of three heat-transfer processes – conduction, convection and radiation. These processes are affected by:

Temperature gradient, which is a temperature difference across the thickness of a wall, roof, floor or other envelope element.

Internal **thermal resistance** of the material to **conduction** of heat flow under steady state conditions.

Convection, the time rate of heat exchange of a material's surfaces with the atmosphere (such as a wall losing heat as wind moves across it).

Radiant energy, both of heat and light (such as heat gained by a wall with sunlight directed on it).

Thermal capacity, the quantity of heat absorbed by – or added to – a material per unit of temperature rise under transient conditions (such as the amount of heat that is held over time as a material gains or loses heat).

The **temperature gradient** or difference varies constantly with movement of the sun, weather changes and indoor temperature. If the temperature difference were held constant over a long time, "steady state" **thermal resistance (R)** values would apply. A steady state calculation of heat flow is satisfactory when the time required for heat to travel through a wall is short compared to the

Integration of Comfort Needs and Natural Energy Flows

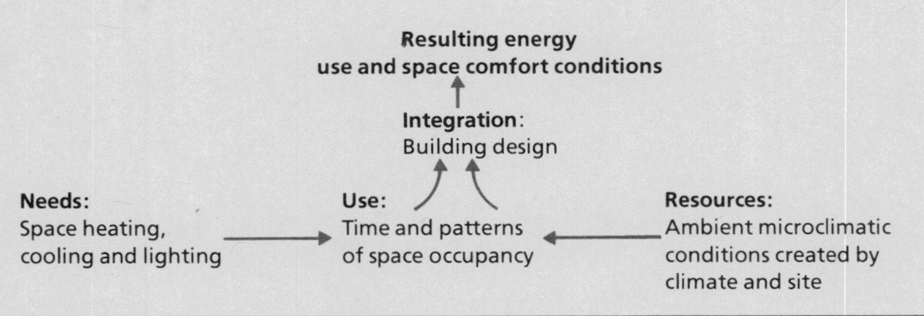

Figure 1-4

time over which outside conditions are changing. Thus, a lightweight thin metal curtain wall which conducts heat into a space in minutes, or a heavyweight wall exposed to a prolonged period of cold lasting several days, will both produce near steady state conditions.

Since heat flow through a wall is a function of the reciprocal of the wall's resistance (U = 1/R), **increasing the resistance by adding equal amounts of insulation results in diminishing savings, as seen in figure 1-5.** For an insulating core of "glass, wood, cotton fibers," as circled in the figure, increasing insulation from 1" to 2" reduces U by 0.09. Adding another inch reduces U further by only another 0.04.

Any **insulation has the effect of widening the spread between temperatures at the warm and cold sides of a material.** That is, the cold side will be colder and the warm side warmer than if there is no insulation. Furthermore, if the wall is exposed to direct sunlight with insulation located on the exterior, as shown in **figure 1-6B,** greater extremes of temperature will occur on the exterior surface than if the insulation were placed on the interior as in **figure 1-6A.** In addition, the exterior wall material will change temperature faster, as there is less thermal capacity outside of the insulation to delay the rise in temperature.

To resist or promote heat flow, it helps to see how heat flows through a building envelope.

Heat may travel by convection, which occurs when air or water molecules move and carry heat with them past surfaces. The contact transfers heat between them. An example is hot air rising inside a chimney and heating up its walls.

The convective quality of a material is **surface conductance,** which is determined by roughness of the surface and air movement over it. **Figure 1-7** shows an increasing rate of surface conductance due to faster air velocity and rougher surface textures. It has little effect compared with insulation in building construction.

Conduction occurs when molecules transfer heat through physical contact from one to the next. They collide as their movement increases due to heat. The effect of conduction is often overlooked in the design of curtain wall systems and windows as heat flow through the frame is ignored. The best way to stop conducted heat is a thermal break of low conductance.

Radiant energy consists of electromagnetic waves which are transmitted through space like buckshot. When

Transmission Coefficient U: Light Weight Prefabricated Curtain Type Walls*

For Summer and Winter
BTU/(hr.)(sq. ft.)(deg. F temp. diff.)

All numbers in parentheses indicate weight per sq. ft. Total weight per sq. ft. is sum of wall and finishes.

Facing
Core Material

Δ = .02
Δ = .04
Δ = .09

Insulating Core Material	Density (lb./cu. ft.)	Metal Facing (3) Core Thickness (in.) 1	2	3	4	Metal Facing with 1/4" air space (3) Core Thickness (in.) 1	2	3	4
Glass, Wood, Cotton Fibers	3	.21	.12	.08	.06	.19	.11	.08	.06
Paper Honeycomb	5	.39	.23	.17	.13	.32	.20	.15	.12
Paper Honeycomb with Perlite Fill, Foamglas	9	.29	.17	.12	.09	.25	.15	.11	.09
Fiberboard	15	.36	.21	.15	.12	.29	.19	.14	.11
Wood Shredded (Cemented in Preformed Slabs)	22	.31	.18	.13	.10	.25	.16	.12	.09
Expanded Vermiculite	7	.34	.20	.14	.11	.28	.18	.13	.10
	20	.44	.27	.19	.15	.35	.23	.18	.14
Vermiculite or	30	.51	.32	.24	.19	.39	.27	.21	.17
Perlite Concrete	40	.58	.38	.29	.23	.43	.31	.25	.20
	60	.69	.49	.38	.31	.49	.38	.31	.26

Equations: Heat Gain, Btu/hr = (Area, sq. ft.) × (U value) × (equivalent temp. diff. Table 19)
Heat Loss, Btu/hr = (Area, sq. ft.) × (U value) × (outdoor temp. − inside temp.)

*For addition of insulation and air spaces to walls, refer to Table 31, page 75.

**Total weight per sq. ft. $= \dfrac{\text{core density} \times \text{core thickness}}{12} + 3$ lb/sq. ft.

Figure 1-5 *Source: Handbook of Air Conditioning Systems Design, Copyright 1965 McGraw-Hill, Inc., Reproduced with permission of publisher.*

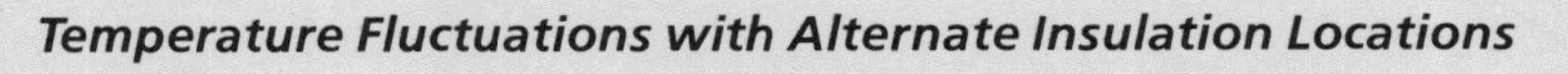

Temperature Fluctuations with Alternate Insulation Locations

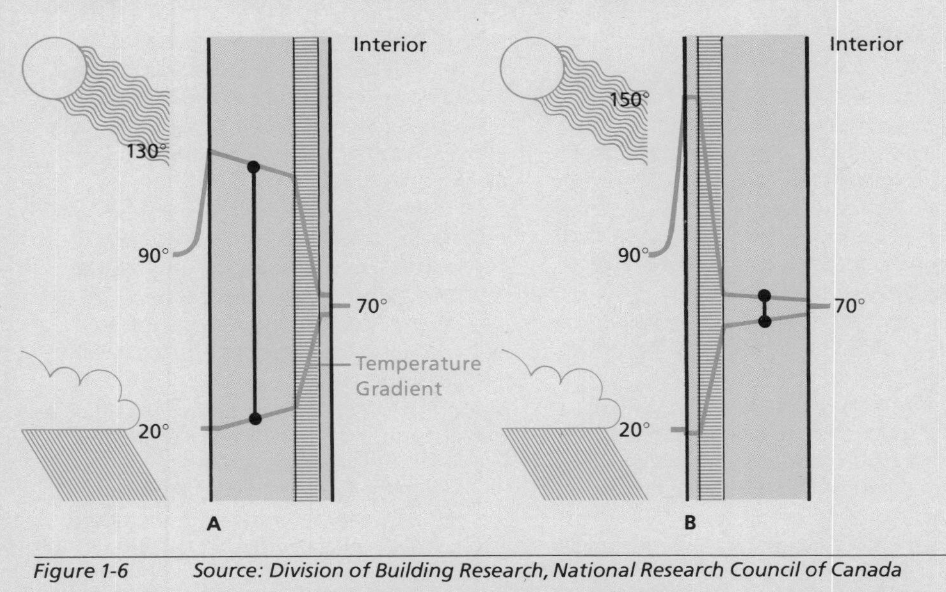

Figure 1-6 *Source: Division of Building Research, National Research Council of Canada*

Surface Conductance for Different 12 in. square Surfaces as Affected by Air Movement

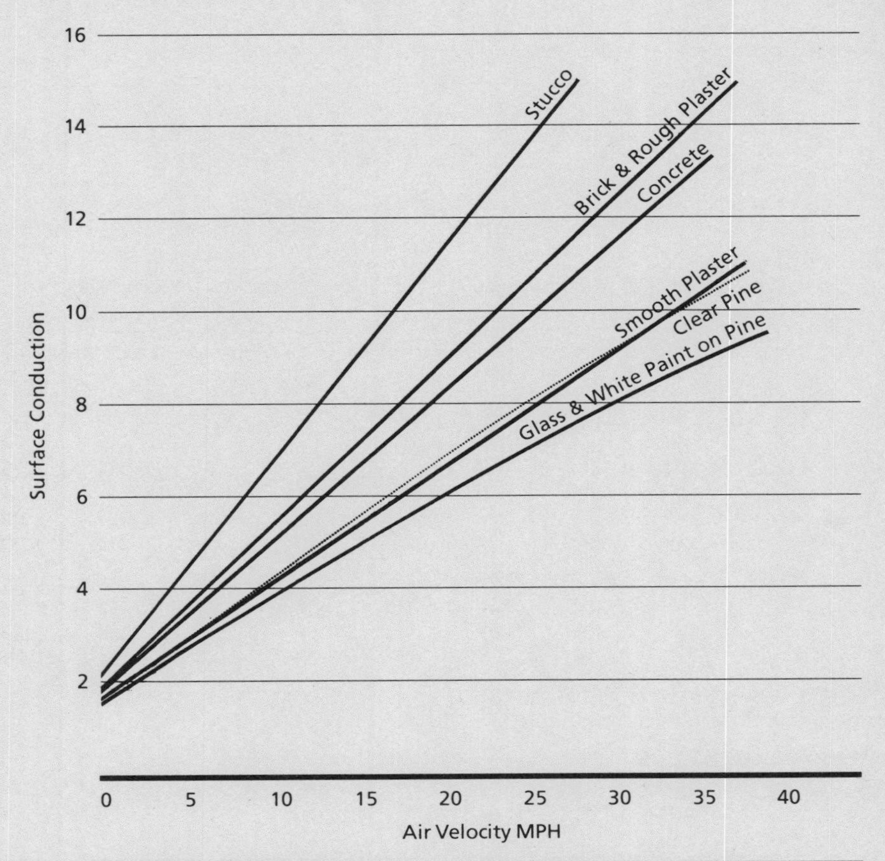

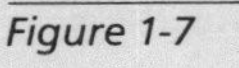

Figure 1-7 *Reprinted with permission from 1977 Fundamentals Volume, ASHRAE Handbook and Product Directory*

they are absorbed by a material, that material heats up. All substances radiate energy to and from others, as a function of absolute temperature difference to the fourth power. The unclothed human body is a very good absorber and radiator of radiant energy. It absorbs solar and other radiant heat, but radiates heat to surfaces colder than the skin. Thermostats do not sense radiant energy as accurately as the body. Therefore, radiant effects are often overlooked in the design of buildings, even though they can be both helpful and harmful.

Excessive radiant heat can be reduced by lowering the temperature of the radiant body, or by blocking the radiant energy by means of layers of thin, smooth, highly reflective materials such as coatings on reflective glass or foil accompanied by an air space. Reflective surfaces may be used to direct radiant energy into more profitable channels.

The **heat storage or thermal capacity** of the materials in the building envelope determines how much energy is required for a wall to change temperature. Thermal storage serves to delay and dampen exterior natural flows.

Because building envelope assemblies neither heat up nor cool down instantly, ignoring their storage capacity can lead to an overestimating of heat flows.

The effect on energy flows in a space due to thermal capacity is determined chiefly by:

- Specific heat of the wall materials.
- Density of wall materials and their placement.
- Local climatic fluctuations in temperature and solar gain.
- Surface area exposed to the elements.
- Interior temperature fluctuations.
- Internal heat generation.
- Solar absorptive characteristics of the material's surfaces.

A net reduction in energy flow through a wall always results when thermal capacity is included in a dynamic analysis. For a combination of local weather and internal room conditions, walls with low insulating value but high thermal inertia may have the same effect on energy consumption as walls with high insulating value and low thermal inertia.

A building envelope for a mild climate and/or an occupancy that generates high internal heat (such as an office or computer facility) is a special design challenge. **Although insulation helps retain heat in cold climates, it may slow dissipation of excess heat in buildings which produce more heat than needed to replace losses to the outdoors.** If an overabundance of internal heat is more predominant than a need for heat, then the amount of insulation may best be reduced. For a given climate and building this "trade-off" must be carefully considered.

In some climates, thermal inertia in the building envelope has little impact on accuracy of the calculation of winter heat loss, but it must be included in the summer heat gain calculation for the same wall to avoid severe inaccuracies.

Figure 1-8 indicates the effect of **thermal storage capacity within the internal building structure** on a building's dynamic energy flows when light fixtures are turned off. The building structure includes floors, elevator cores, and contents. Less than half of lighting heat generally appears as an air conditioning load instantaneously because the radiant heat from the light is stored by the structure. After the lights are turned off, this stored heat reappears as a delayed load on the cooling system. If the air conditioning is turned off with the lights as shown, all the stored heat except that small amount which is dissipated during the night must be removed as a start-up load the next morning.

As the mass of the building increases, the release of this load stretches out further. A similar curve occurs for solar heat gain. The heavier the building, the more the impact is delayed and the peaks reduced as shown in **figure 1-9.**

The sun's energy impact on the building envelope depends on how well the envelope is designed.

The sun is in a different position in the sky every moment of the year, but its position is predictable. The energy planner can therefore easily orient the building and its openings to receive the sun, while rejecting it when not wanted. Thus, the sun warms in winter from a low altitude, and overheats in summer from a higher altitude. In the northern hemisphere it seldom shines on the north except in early morning or late evening in summer. The impact varies drastically by orientation, season and hour. A sample of the *ASHRAE Handbook of Fundamentals,* Solar Heat Gain Factors illustrates these forces clearly and is included as **figure 1-12.**

Equivalent Temperature Difference (ETD) combines the effect of outside to inside temperature difference, solar effect, and heat storage into a

Actual Cooling Load from Fluorescent Lights Average Construction

Actual Cooling Load, Solar Heat Gain, Light, Medium and Heavy Construction

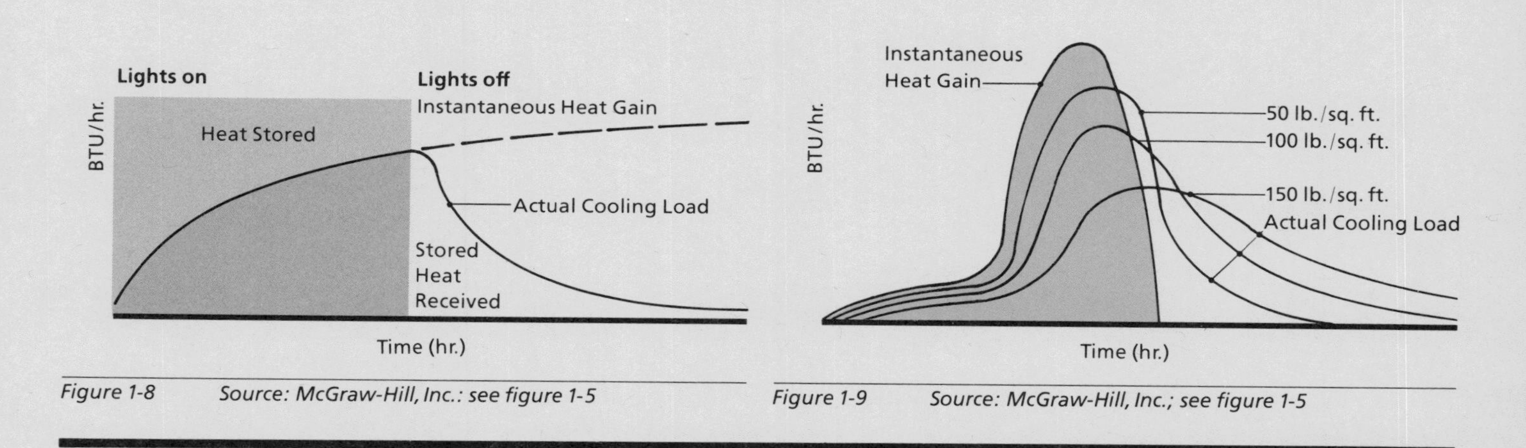

Figure 1-8 *Source: McGraw-Hill, Inc.: see figure 1-5*

Figure 1-9 *Source: McGraw-Hill, Inc.; see figure 1-5*

Equivalent Temperature Difference (°F)

Based on 95°F db Outdoor Design Temp; Constant 80°F db Room Temp; 20°F Daily Range; 24-hour Operation; July and 40° N. Lat. **Exposed to sun**

Weight of Roof (lb/sq. ft.)	**Sun Time** AM 6	7	8	9	10	11	12	PM 1	2	3	4	5	6	7	8	9	10	11	12	AM 1	2	3	4	5
10	−4	−6	−7	−5	−1	7	15	24	32	36	43	46	45	41	35	26	22	16	10	7	3	1	−1	−3
20	0	−1	−2	−1	7	9	16	23	30	36	41	43	43	40	35	30	25	20	15	12	8	6	4	2
40	4	3	2	3	6	10	16	23	28	33	36	40	41	39	35	32	28	24	20	17	13	11	9	6
60	9	8	6	7	8	11	16	22	27	31	35	38	39	38	36	34	31	28	25	22	18	16	13	11
80	13	12	11	11	12	13	16	22	24	28	32	35	37	37	35	34	34	32	30	27	23	20	18	14

Figure 1-10 *Source: McGraw-Hill, Inc.; see figure 1-5*

single factor and is useful in calculating heat gain through walls and roofs. **Figure 1-10** is a sample of data used to determine ETD for a given room and latitude under several conditions.

The sun's impact is related to the solar absorptive surface characteristics of the materials, as seen in **figure 1-11.**

Solar heat gain factors measure solar heat passing through double strength ⅛" sheet glass under clear conditions at specific latitudes at sea level. They can be modified for local conditions of clearness, cloud cover, and air moisture to improve accuracy. Separation into diffuse and direct insolation is advised as is use of on-site recorded data. As this is seldom available, the energy planner should relate the level of solar data needed for the project to the sensitivity of the analysis.

Reflectivities and Emissivities of Some Typical Surfaces

Surface	Reflectivity percentage to solar radiation	Reflectivity percentage to thermal radiation	Emissivity percentage of thermal radiation
Silver, polished	93	98	2
Aluminum, polished	85	92	8
Copper, polished	75	85	15
White lead paint	75	5	95
Chromium plate	72	80	20
White cardboard or paper	60-70	5	95
Light green paint	50	5	95
Aluminum paint	45	45	55
Wood, pine	40	5	95
Brick, various colors	23-48	5	95
Gray paint	25	5	95
Black matte	3	5	95

Note: There is often considerable variation from one sample of material to another. The figures given above are to be taken as illustrative, not as means of numerous samples.

Figure 1-11 *Found in Physiological Objectives in Hot Weather Housing, an introduction to hot weather housing design; US Department of Housing and Urban Development, Office of International Affairs.*

Sample of Heat Gain through 1 sq. ft. of Glass

Air-Conditioning Cooling Load

Solar Position and Intensity; Solar Heat Gain Factors for 32 Degree North Latitude

Date	Solar Time A.M.	Solar Position Alt.	Solar Position Azimuth	Direct Normal Irradiation, Btuh/sq. ft.	N	NE	E	SE	S	SW	W	NW	Hor.	Solar Time P.M.
Jan 31	7	1.4	65.2	1	0	0	1	1	0	0	0	0	0	5
	8	12.5	56.5	202	8	29	160	189	103	9	8	8	32	4
	9	22.5	46.0	269	15	16	175	246	169	16	15	15	88	3
	10	30.6	33.1	295	19	20	135	249	212	45	19	19	136	2
	11	36.1	17.5	306	22	22	67	221	238	110	22	22	166	1
	12	38.0	0.0	309	23	23	25	174	246	174	25	23	176	12
				Half Day Totals	**75**	**91**	**529**	**974**	**834**	**262**	**75**	**75**	**509**	
Feb 21	7	6.7	72.8	111	4	47	102	95	26	4	4	4	9	5
	8	18.5	63.8	244	12	64	205	217	95	12	12	12	63	4
	9	20.3	62.8	287	19	32	199	248	149	19	19	19	127	3
	10	38.5	38.9	305	23	24	151	241	189	31	23	23	176	2
	11	44.9	21.0	314	26	26	76	208	213	87	26	26	207	1
	12	47.2	0.0	316	27	27	29	155	221	155	29	27	217	12
				Half Day Totals	**97**	**207**	**749**	**1091**	**780**	**227**	**98**	**97**	**689**	
Mar 21	7	12.7	81.9	184	9	105	176	142	19	9	9	9	31	5
	8	25.1	73.0	260	17	107	227	209	62	17	17	17	99	4
	9	36.8	62.1	289	23	64	210	227	107	23	23	23	163	3
	10	47.3	47.5	304	27	30	158	215	144	29	27	27	211	2
	11	55.0	26.8	310	30	31	82	179	168	58	30	30	242	1
	12	58.0	0.0	312	31	31	33	122	176	122	33	31	252	12

(Columns N through Hor.: Solar Heat Gain Factors, Btuh/sq. ft.)

Figure 1-12 *Source: ASHRAE Handbook; see figure 1-7*

Figure 1-12 illustrates one day of summer heat gain through a common size window if it were facing eight ways or horizontally. Heat gain due to solar radiation is many times greater than the gain conducted through the single-glazing due to indoor-outdoor temperature difference. The amount saved by double-glazing would be about half that conducted. It would be small compared to heat from radiation. For big summer savings, it is better to reflect the sun away from the glass, either through shading devices or reflective glass surfaces, than it is to use double-glazing.

The timing of solar gain is important. In **figure 1-13**, the west sun is significant for an air-conditioning peak because it comes after the building has stored up most of its heat and the outside air being brought into the building is at its hottest. The envelope is a device for delaying and reducing these coincident peaks.

Shading coefficients and descriptions of tests are available from window manufacturers. Coincident glass temperatures from these tests should be requested to be able to judge their effect on the mean radiant temperature experienced by occupants. (Egan: *Thermal Comfort*, Ref.)

Cooling Load for 100 sq. ft. Double Glazing on a Typical Clear Summer Day (45° latitude, shading coefficient = 0.55)

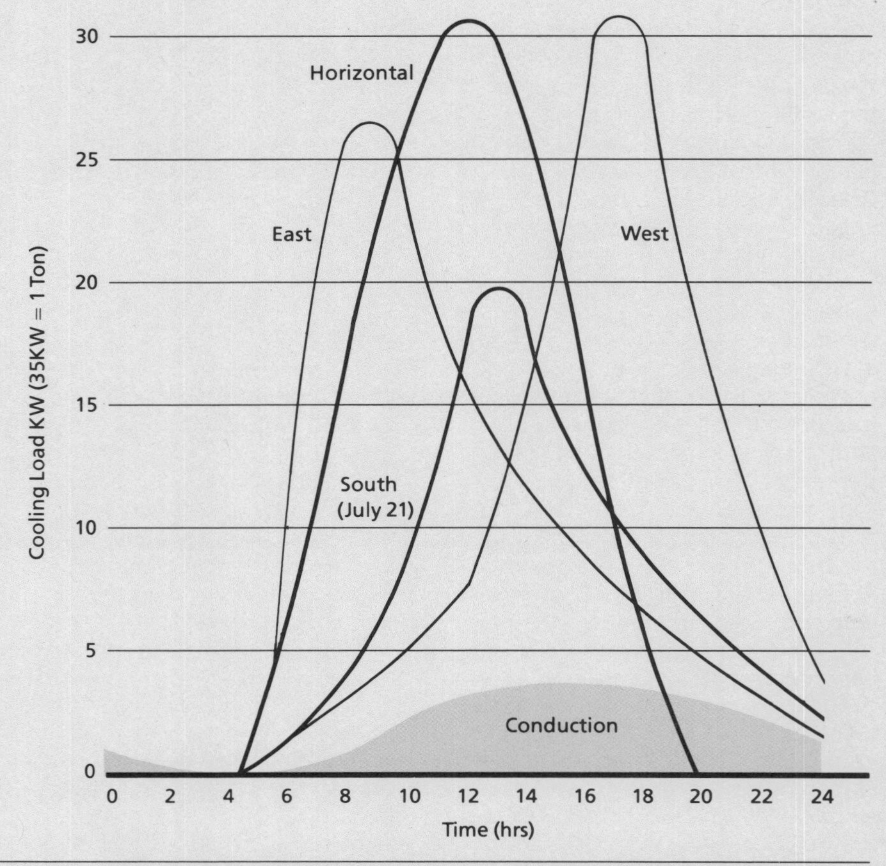

Figure 1-13 *Source: Division of Building Research, National Research Council of Canada*

Ventilation

Supplying and removing air from a space to maintain an acceptable level of air quality is a prerequisite for both health and comfort. Once a minimum acceptable level is reached, however, added amounts are of little value unless the energy planner is concerned with a second use of ventilation, which is to move air across the skin to help carry off body heat. Since the heating or cooling of ventilation air can consume as much as 25% of a building's HVAC energy in cold or warm humid climates, too much outside fresh ventilation air is to be avoided. This calls for attention both to the intentional (mechanically controlled) and unintentional (infiltration) outside air quantities brought into a building.

Ventilation provides enough oxygen for breathing, removes carbon dioxide generated in the breathing process and prevents build-up of odors. The first two requirements are related to the number of people present and may be met by supplying as little as 5 CFM per person of outside air.

Odor control requirements depend on the source, magnitude and type of odors generated in a space. Controlling tobacco smoke requires some 15 to 25 CFM of outside air per person smoking, depending on the acceptable room smoke level. If ventilation requirements for odor control are much above the requirements for breathing, the air filtration system may be upgraded to provide odor removal, along with the usual removal of dust and lint particles. In an air-conditioned building, reducing ventilation air quantities to the minimum required for breathing will sometimes bring down energy consumption enough to justify the cost of odor removal equipment.

For an area of contamination that can be isolated from conditioned spaces, a second process — **exhaust of conditioned air from conditioned spaces**— may be used. This exhausts air from the immediate vicinity so that contaminants are captured and directed into the exhaust system before they can diffuse into the occupied space. For example, manufacturers of exhaust hoods now provide features that help reduce the amount of conditioned exhausted air by direct use of outside air for makeup in the immediate hood area.

In a building that is not air-conditioned, **air movement** aids evaporation by removing perspiration and maintaining a body's temperature equilibrium. As air moves past occupants, body evaporative heat losses rise, providing additional relief from heat and humidity. Air movement also helps dissipate internal and solar heat gains.

In such cases, ventilation air quantity is usually increased much above that required for odor control, so as to provide relief cooling of a space in hot weather. Air quantities are usually quite large – from 2 to 4 times total air circulated in a comparable air-conditioned building, and from 20 to 50 times the minimum outside air required for breathing. Due to the large outside air quantities required, the ventilation air system should be controlled in a way that will reduce ventilation air to a reasonable minimum whenever outside temperature is cool enough to permit building temperature control with smaller air quantities.

Since the energy required to move these larger air quantities may be substantial, the energy planner should look at natural ventilation methods using openings in the building envelope. **The building must then be designed from the beginning to allow for the desired air flow patterns.** Natural ventilation may be induced by installing fixed or adjustable projections that guide the wind into poorly oriented spaces. References by Olgyay and Givoni deal with such design methods.

An **"economizer cycle"** on an air handling damper system allows an air-conditioning system to use outside air for cooling when it is cooler or dryer than inside air. This helps eliminate the refrigeration that is otherwise needed. Outside air dampers may open completely so that 100% outside air is distributed.

In winter, outside air leaking through openings and cracks in doors, windows, walls and roofs can physically harm both building and occupants. Air leaks in and out whenever there is a pressure difference across a building surface. Air pressure differences are hard to calculate and measure. They may be due to the effects of wind, a difference between indoor and outdoor temperatures and air densities, or the balancing and operation of mechanical ventilation and exhaust systems.

When temperature difference varies proportionally with building height, there is a chimney or **stack effect.**

Exfiltration is the outflow of conditioned air and likewise may cause problems. For example, condensation may occur in hidden parts of walls or roofs. This may lead to structural deterioration. Condensation may be overcome or controlled by ventilating attic areas with outside air, ensuring that air tightness in the inner part of a structure is several times tighter than the outer cladding, and restricting air flow between floors of multi-story buildings so as to reduce large pressure differences resulting from the stack effect.

Viewing Task Silhouetted Against a Glare Source

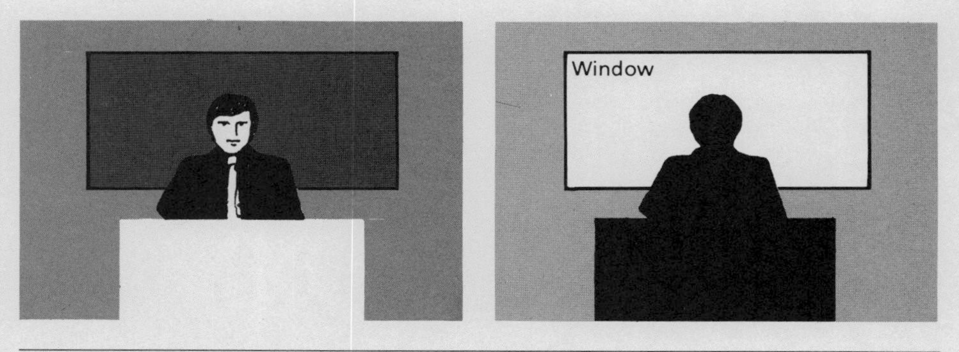

Figure 1-14

As the cost of conditioning outdoor air rises, buildings will gradually be forced to use other ways to remove odor contaminants. These may include **filtering** air streams through beds of activated charcoal pellets to adsorb odors.

Reclaiming waste heat from process activities for reuse with ventilation air is possible through a heat recovery device called a thermal wheel. This works much like an air-to-air heat exchanger. It is used to temper ventilation air and preheat combustion air. Other heat exchangers may be applied to domestic water preheating and terminal reheat.

Lighting

Light communicates the forms, textures, surfaces, voids and colors which stimulate psychological response to a building.

Building occupants require the right amount of illumination to view a specific range of tasks. Older people need more light and contrast than do younger; a faint copy of a lawyer's brief requires more illumination than a clear handwritten document in black felt tip pen. Insufficient light slows down an activity.

Designers of lighting too often supply more than enough quantity of illumination but lose sight of quality. Quality begins with contrast.

The degree of perceived **contrast** in a room is the difference between intensity and amount of bright areas and dark areas within a field of view. A small bright light source can create a pleasant "sparkle" in a room, such as candlelight on a dinner table, yet a series of large bright windows may cause a darker interior to appear gloomy. To counter this effect, electric lighting must be increased on the interior of the space to balance exterior wall brightness (which consumes energy), or else glare must be reduced at the daylight source.

Illumination levels that result once glare is reduced should parallel the requirements of a visual task. For example, a desktop should be illuminated most, the background to the desk next, then other light sources in the field of vision. Large lamps should be out of sight and views of direct sunlight should be controlled as to brightness. By matching task importance, energy efficient lighting design focuses on the viewer's needs, so that lighting quality is improved as energy costs go down.

Glare arises when a high intensity source relative to the visual task comes into the viewer's field of vision. The eye is drawn to the bright source, whereupon the eye muscles adjust the light entering the eye as a camera does, so that darker visual tasks appear even darker **(figure 1-14).**

Veiling reflections occur when a bright light source is reflected off a task into the viewer's eyes. Needed visual contrast is lost. The usual response of an office worker conscious of this problem is to bring in a task lamp. This puts enough light on the task from the right direction to overcome the reflection. Such an alternative consumes additional energy. A better response is to move the existing fixture.

How **color** in lighting shapes our ability to see and respond has been lost with the recent preoccupation with quantities of footcandles. Lightmeters read illumination within a predetermined, weighted selected partial color range which some experts say is inaccurate when compared to the eye.

To quantify color, light sources are visually compared to a standard lamp burning at a "color temperature." As color temperature is reduced and produces warm tints, fewer footcandles are needed for viewing compared with the higher temperatures of the bluish spectrum. Quantified color temperature coincides only vaguely with the way we see, proving again that lighting design cannot be entirely quantified.

Light quality depends on:

glare
brightness (luminance) and brightness ratios
direction
visual clarity (color)

Units of Lighting Measurement

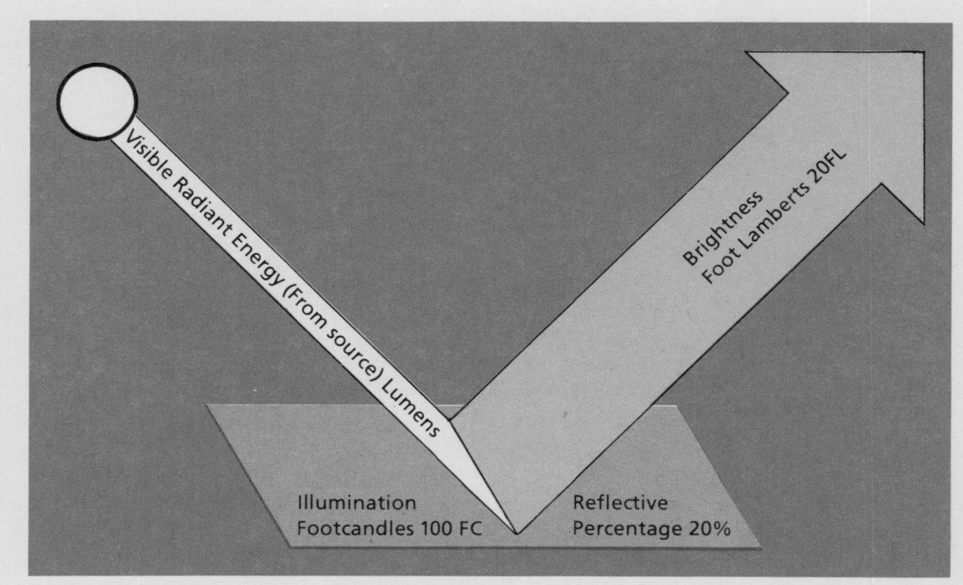

Task reflectance x Illumination = Task Luminance
(per cent) (footcandles) (foot lamberts)

Figure 1-15

Lower reflectance and difficult tasks call for higher illumination.

Measurable lighting levels are suggested by the *Illuminating Engineering Society Handbook* (Ref). Units of measurement are shown in **figure 1-15.** Because of energy concerns in many states that are reevaluating standards, the designer may find that lower levels are both adequate and acceptable to clients. These levels are directed to the task, and are intended to be acceptable to statistical groups of people under statistically normal conditions. This works well so long as orientation of occupants, position of light sources, access to daylight, reflectance of surfaces, hours of occupancy and position of daylight sources are all synchronized with the visual needs of building occupants' activities and locations. Works by Stein (Ref.) and G.S.A. (Ref.) are useful in establishing lighting levels.

The best way to **predict how well a lighting design will work** and to check calculations is to build it, install the fixtures and other devices, and live with it. Since this is seldom practical, the next best approach is to come as close to construction as possible – by building a full-size mock-up or a small model, then installing the artificial light source at the site, in the model or in the energy planner's office. Be sure the client experiences the effect through viewing the model or photographic slides of the results.

Large organizations often require large clerical staffs. Over the years this has resulted in multi-story deep office spaces isolated from windows. Combined with cheap electricity, it served to do away with interest in **daylighting.** Buildings tended to be designed with lighting as a remote concern rather than as a primary form-giver. Instead of using artificial lighting as a substitute when daylighting was not enough, design professionals eschewed daylighting because of its potential as a source of unwanted heat. Since electrical lighting has commonly accounted for 50% of energy costs in office buildings in temperate climates, one must, as electricity costs increase, take a fresh look at the economics of daylighting.

Although soundly designed daylighting has many good points, the **accompanying solar heat gain** can be a drawback. For example, the lumens in a single square foot of direct clear winter sunlight can theoretically provide 50 footcandles (FC) of illumination over an area of 180 sq. ft. if uniformly distributed. The same square foot of daylight will contribute over 200 BTU/hr of heat to the space, which may be good or bad depending on the building's capacity to use the heat. These forces are powerful and they can vary rapidly, so that good design is critical.

In some cases, the cost of meeting the many special conditions required for daylighting may be better invested in a good artificial lighting system. Energy efficient building design is always a matter of trade-offs.

In daylight design first decide to what extent the design should make use of **diffuse and/or direct daylighting,** based on user activities and the climate.

Major Variables in Daylighting Design

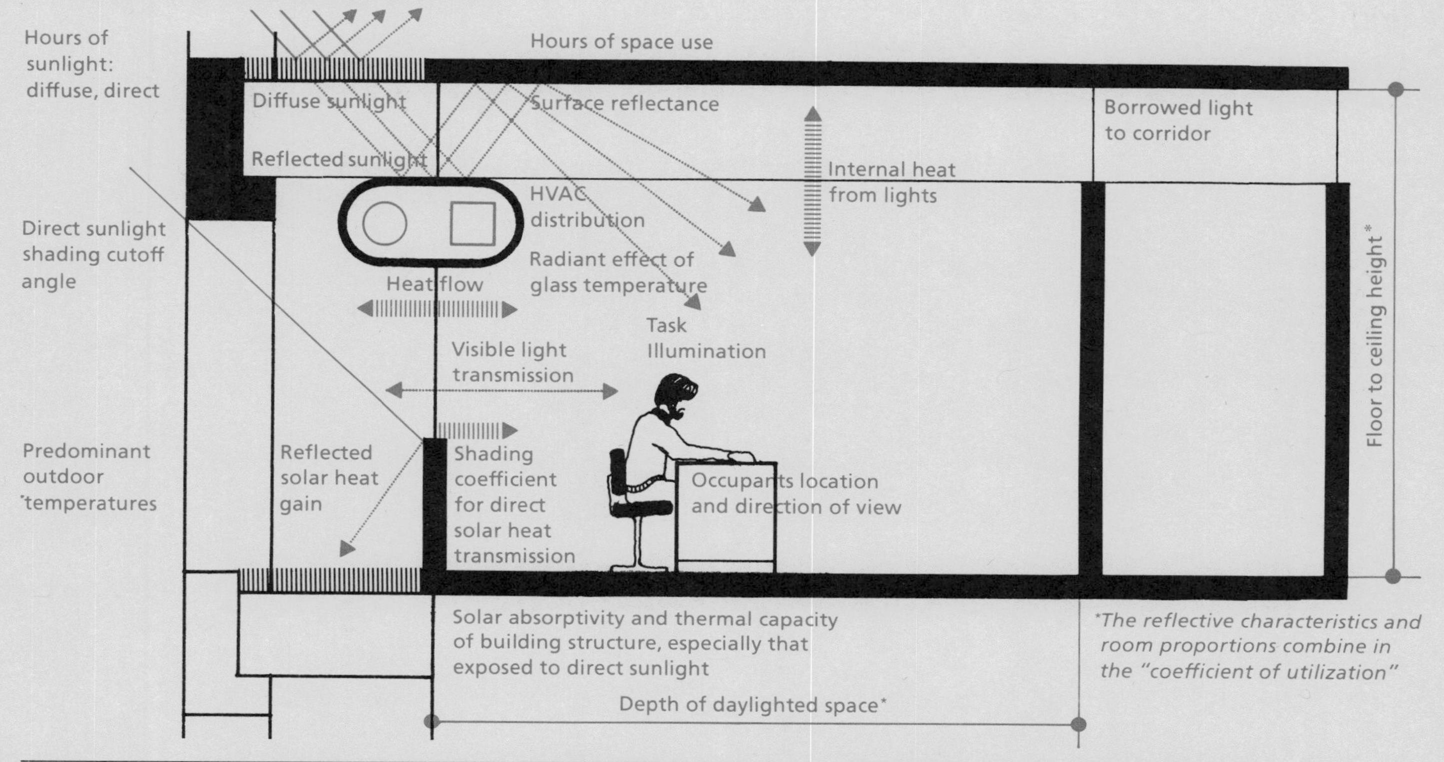

Figure 1-16 *Source: Shell Woodcreek Exploration and Production Offices, Houston, Texas; CRS Architects and Engineers*

Diffuse daylight:

Results from north light or overcast, when sunlight is scattered and reflected by atmosphere and clouds.

Does not have as intense heat or brightness as direct sunlight; hence large windows are desirable.

Is much easier to use and control than direct daylight.

Decays quickly as the task moves farther from the window; it is generally useful in a space only about twice the height of the window.

Is non-directional, does not depend on orientation, and cannot be precisely controlled or reflected.

Direct daylight:

Can be reflected over large distances if the reflector position can be frequently adjusted through controls.

Occurs under clear skies.

Is intense, so that small openings may be adequate – although the brightness ratio may become undesirable.

Is accompanied by intense heat.

Can provide visual "sparkle."

Often requires shading or glare reduction.

The next decision in daylighting concerns **heat.** This is based on the balance between heating and cooling needs and the time and location of solar heat. One would not orient glass horizontally to the maximum summer sun in a hot climate to provide daylighting. The more likely solution is shaded to the south, or an orientation towards the north.

The **key variables of daylighting** are listed in **figure 1-16.** They are modified and combined to create the most effective lighting and thermal balance.

For example, a window may be divided into a low vision panel, with glare-reducing, solar heat-reflecting glass, while an upper panel of clear glass with a reflector device throws light into the interior. In this case, ceiling height would need to be increased beyond the standard. The HVAC distribution system must be able to pick up excess solar heat on the south side, and deliver it to the north side; or else a thermal storage device is required, such as a massive solar absorptive surface which would collect excess heat for release later.

Light **interior surface reflectances** and artificial lights would balance the light on the darker interior wall with the brightness of sky and ceiling. Switching of artificial lights would be keyed to space use, and daylight controlled automatically or by occupants. The IES (Ref.) and Griffith (Ref.) offer more information.

The yearly cumulative balance between hourly heat loss, internal heat generation and solar heat gain into the room is critical to daylighting design. That is because economic feasibility is determined by the degree to which the daylight and its heat are helpful rather than harmful. By and large, when increasing glass area in an office with high internal heat generation, one should consider these factors:

Glass allows daylight to replace electrical lights operation, and heat accompanying daylight should be less than the heat from the fluorescent fixtures which can be turned off.

In moderate to warm climates, glass may allow high internal overheating to be more easily dissipated to the outside during overheated periods. Only an hour-by-hour calculation will show whether glass area should be **increased** to dissipate heat, or **reduced** to hold it in.

As a glass area is increased beyond the optimum for daylighting, the added daylight provides less and less benefit. Meanwhile, the additional solar heat may overheat the space.

As a result, **there is a best window size for a given climate, design and occupancy.**

Reducing electrical lighting offers big opportunities for cutting back on energy use, so long as certain conditions are observed:

Space heating costs may go up and offset some apparent savings as beneficial heat from lights is reduced.
In a redesign, the HVAC distribution system may need to be rebalanced to avoid overcooling after lights are removed.
The room may appear gloomy.
New switching may be required.
Removing lights may meet with user resistance if not explained.
Electricity use will drop.
Space cooling cost will go down as heat from lights is reduced.
If daylighting is used, passive solar gain may provide useful space heating.
Seeing may become more comfortable.
Longer lamp life leads to less relamping and maintenance.
Air-conditioning peak load will come down, so air-conditioning equipment capacity and/or operating time can be reduced too, with resulting less wear on equipment.
In mild climates available daylighting usually peaks at a time of greatest cooling demand. Hence, daylighting is likely to reduce a building's peak electric demand.

As its most important benefit, sunlight not only contains a full range of visible colors, but lets occupants' **psychological responses** shift with its colors, direction, and intensity throughout the day and year. A room admitting blue north winter light makes for a different atmosphere from that in the same room at a warm red summer sunset or in the firelight of a hearth.

Therefore, as daylight changes over time, it can also bring occupants' "inner clocks" into harmony with their environment and create a fine sense of well-being.

Energy sources

Energy is usually defined as the capacity for doing work. More generally, it is the ability to produce an effect. To fulfill needs in buildings, energy is transformed from one form into another. The important forms of energy in planning work include thermal, mechanical and electrical.

Although each form has its own units, the standard form of thermal energy measurement in buildings is the **British Thermal Unit (BTU).** This is defined as the heat required to raise the temperature of 1 lb. of water by 1°F. Multiples of BTU's are used when large numbers are involved, for example:

1 MBTU = 1,000 BTU
1 therm = 100,000 BTU
1 Barrel of Oil Equivalent (BOE) = 5,800,000 BTU

In studying a building's energy use, the energy planner is concerned with the total amount of energy used (consumption) and the rate at which energy is used during a time period (energy demand).

Energy consumption = Energy demand × time
Energy demand = Energy consumption ÷ time

The BTU is thus a unit of energy consumption; the BTU per hour (BTUH) is the corresponding unit of energy demand, and refers to the number of BTU's used per hour.

The units commonly used to express **power** are the kilowatt (electrical power) and the horsepower (mechanical power). Power is the rate at which energy is used and is sometimes referred to as demand.

Figure 1-17 summarizes these relationships.

This book concentrates on the energy transformations which take place within the building, so that one may determine how much energy must be supplied to the **building line.** The energy required to extract resources, to convert and distribute them is beyond the scope of the book (for an analysis of energy as related to resource impact and use, see Section 12, **Addendum to ASHRAE Standard 90-75** (Ref.)).

The conversion factors listed earlier can be used to express rough equivalents of different energy forms. For instance,

1 KWH of electricity = 3,413 BTU of heat.

Thus, for an electric resistance heater to produce 3,413 BTU of heat will require 1 KWH of electricity delivered to the building line plus a small loss in the building wiring.

This is the simplest case. Since the conversion of electricity to heat by means of electric resistance is close to 100% efficient, **direct conversion** may be used.

In other cases, energy is lost through conversion, and must be considered. For example, losses may result from heat of friction in the motor or engine doing the converting.

Energy planners must note that 1 KWH of electricity = 1.34 HP-HR of **mechanical work.** But in an electric motor in which 10% of the input electric energy is lost to forms of energy other than work – such as heat of friction – the conversion becomes:

1 KWH input to the motor = 90% × 1.34 HP-HR of work output of the motor.

Efficiency is the ratio of output energy to input energy:

$$\text{Efficiency} = \frac{\text{Energy output}}{\text{Energy input}}$$

Therefore, if we need 1.34 HP-HR of work and motor efficiency is 0.9, 1.11 KWH of electricity will be required.

$$\text{Required KWH (input)} = \frac{1.34 \text{ HP-HR (required output)} \times 1 \text{ KWH}/1.34 \text{ HP-HR}}{0.9 \text{ (efficiency)}} = 1.11 \text{ KWH}$$

There may be further conversion losses (notably heat of friction) in the **distribution** of energy. In the case of electrical distribution systems in buildings, this loss is so small it can be ignored. In the distribution of heat by means of hot air, hot water, or steam, a factor may, however, be included to account for this distribution loss.

Energy estimates may be complicated by the fact that equipment may not operate continuously at uniform power input. In such cases, the average

Acronyms for Energy Forms

Energy Form	Units of Energy Consumption		Units of Energy Demand	
	f.p.s. system	m.k.s. or s.i. system	f.p.s. system	m.k.s. or s.i. system
Thermal	**BTU** British thermal unit	**KCAL** Kilocalorie (1KCAL = 3,968 BTU)	**BTUH** BTU's per hour	**KCAL/SEC**
Mechanical	**HP-HR** Horsepower-hour (1 HP – hr = 2545 BTU)	**JOULE** (1 joule = .000948 BTU)	**HP** Horsepower	**JOULE/SEC** 1HP = 745 J/Sec
Electrical		**KWH** Kilowatt-hour (1 KWH = 3413 BTU)		**KW** Kilowatt (1 KW = 1000 watts)

Figure 1-17

operating rate may be estimated by determining either a usage factor or an average **load factor.** If the equipment cycles on and off, but runs at uniform power while operating, it is best to install a timer to record the actual operating time over a period of several days or weeks. An alternative is to observe operation and record the length of "on" and "off" cycles. If the cycles are plotted, they make a graph as shown in **figure 1-18,** and the usage factor is calculated:

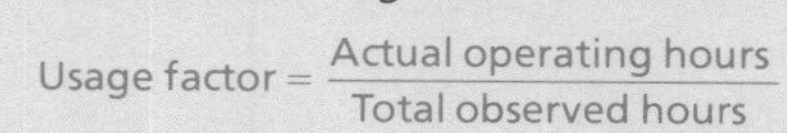

$$\text{Usage factor} = \frac{\text{Actual operating hours}}{\text{Total observed hours}}$$

If equipment runs continuously but power input varies, one can determine the average load factor. One approach is to install a recording watt-meter to record power input. This graph resembles **figure 1-19,** and the average load factor is estimated by calculating the area under the curve and dividing by the operating time.

An alternative is to install a watt-hour meter, which records total energy consumption over the time period. In either case, the average load factor is:

$$\text{Average load factor} = \frac{\text{Sum of (actual power} \times \text{hrs.) over total period}}{\text{Peak design power} \times \text{total hours}}$$

Actual energy consumption for the energy estimate is:

$$\text{Energy consumed} = \text{Maximum power} \times \text{hours} \times \text{average load factor or usage factor}$$
$$(\text{KWH}) = (\text{KW}) \times (\text{HRS}) \times (\%)$$

If **figures 1-18 or 1-19** were plotted for an air-conditioning compressor with a full load power input of 50 KW, actual energy consumption during a 7:30 AM to 3:00 PM time period is:

Equipment Cycles On and Off

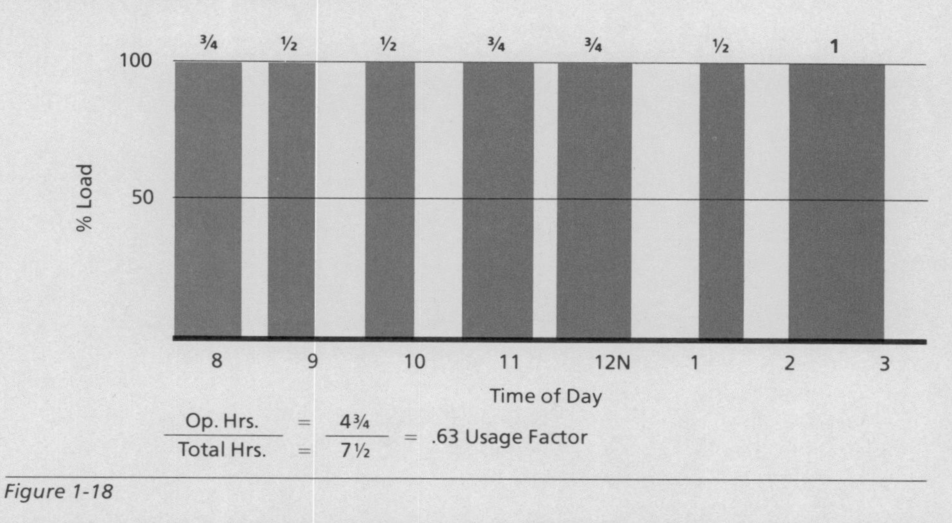

Figure 1-18

Equipment Operates Continuously, but Power Input Varies

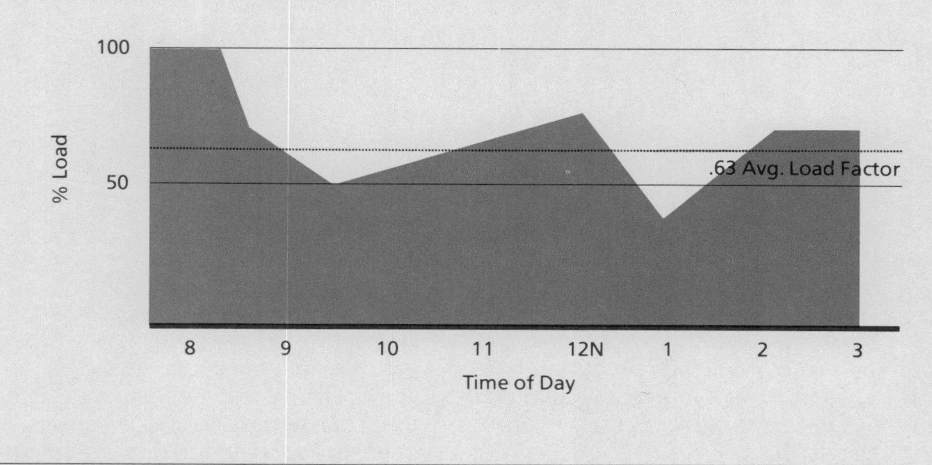

Figure 1-19

$$\text{Energy consumed} = 50\ \text{KW} \times 7\tfrac{1}{2}\ \text{HRS} \times 0.63\ \text{factor} = \text{KWH}$$

Most lights are rated in terms of watts of power input, and the heat output is:

$$\text{Heat output (BTU/HR)} = \text{Power input (watts)} \times 3.41$$

Power used to perform work is usually expressed in terms of BHP (brake horsepower). The efficiency of a machine is the relationship between output and input BHP. This way input work can be computed by finding how much actual work is being done, and what its efficiency is.

$$\text{Machine input BHP} = \frac{\text{Work output BHP (from load and energy requirement)}}{\text{Machine efficiency (from machine ratings)}}$$

Machines are connected to a driving engine or motor by a belt or coupling. If a belt drive is used, it may waste 3% to 5% of the energy it transmits, so the motor output BHP must be greater by this amount. If a fluid type coupling is used, the losses are greater. The motor load (output BHP) is simply the machine load plus any drive losses:

$$\text{Motor output BHP} = \text{Machine input BHP} + \text{drive loss BHP.}$$

An essential concept to understand is that the actual **motor load** (power output) will balance itself to the actual equipment power requirement (equipment load) at that moment. The **motor size** (HP) does *not* tell you the actual motor load (BHP) but only the *maximum* load the motor was designed to handle. A 30 HP fan motor will frequently be operating at 18 BHP load at one time and 25 at another.

Since electric motors supply the power to drive most mechanical equipment, it is useful to be familiar with their performance characteristics. The input power (KW) is related to the output work (BHP) by the motor efficiency and the energy conversion factor from work energy to electrical energy (0.746).

$$\text{Motor Power Input KW} = \frac{\text{Motor Output BHP} \times 0.746}{\text{Motor Efficiency}}$$

The motor efficiency varies with percent load and also with the design of the motor.

Most **alternating current electric motors** are induction type, in which the voltage and current flow are slightly out of phase. This increases the current flow in the power transmission lines and the electric generating plant. The "power factor" of an induction motor is the ratio of the theoretical current flow if the voltage and current were in phase to the actual current flow:

$$\text{Power Factor} = \frac{\text{Theoretical In-Phase Power Flow (Amps)}}{\text{Actual Current Flow (Amps)}}$$

The cost of generating and transmitting power is increased if greater current flow is required because of low power factor, so most commercial power meters record power factor (PF) in addition to electric consumption (KWH). The power cost is increased as the power factor decreases. Power factor can be improved or controlled by avoiding motor operation at low loads, purchase of specially designed motors with superior power factor characteristics, or by installation of capacitors on the electric distribution system to offset the lagging power factor of the induction motors.

Motors 1 HP and larger in size are usually 3-phase type. The power input to **3-phase motors** at any moment can be calculated if you know the power factor and efficiency and measure the voltage and amperage:

$$\text{KW} = \frac{\text{Volts} \times \text{Amps} \times \text{Power Factor} \times 1.732}{1000}$$

$$\text{BHP} = \frac{\text{Volts} \times \text{Amps} \times \text{Power Factor} \times 1.732 \times \text{Motor Efficiency}}{746}$$

Small motors are usually **single phase** type. The power input to these motors is:

$$\text{KW} = \frac{\text{Volts} \times \text{Amps} \times \text{Power Factor}}{1{,}000}$$

$$\text{BHP} = \frac{\text{Volts} \times \text{Amps} \times \text{Power Factor} \times \text{Motor Efficiency}}{746}$$

Engines are sometimes used instead of electric motors to drive machinery. The engine power output is calculated in the same manner as for an electric motor. The input power is usually expressed in terms of the type of fuel used by the engine. The required energy input is determined by the fuel energy conversion factor and the engine efficiency:

$$\text{Engine Power Input (Cubic Ft/Hr Nat. Gas)} = \frac{\text{Engine Output BHP}}{\text{Engine Efficiency}} \times \text{Fuel Energy Conversion Factor}$$

Heating, ventilating and air-conditioning

The HVAC system tends to be the largest single user of energy in a building, so an error in this portion of the energy analysis will strongly affect the accuracy of the total building estimate. Preparing an accurate HVAC estimate is especially difficult because of the many types of HVAC systems and equipment. The energy consumption of two similar buildings with different types of systems or equipment, or even merely different operating schedules and control settings, may vary by as much as 50%.

Air-conditioning is usually defined as controlling the temperature, humidity, ventilation, air motion, and air quality within a building. In ordinary "comfort" applications, emphasis is on temperature control and, aside from building code design ventilation requirements, the other items can vary widely. Humidity, for instance, may change from 70% in summer to as low as 10% in winter. In contrast, some industrial or critical applications call for precise control of all items over the entire operating season. The type of HVAC system and control, and the energy required, obviously depend on the building's usage.

A **typical HVAC system** is shown in **figure 1-20.** In summer, air is drawn from the room through the return air duct and mixed with outside (ventilation) air. The air passes through filters to the cooling coil, where it is cooled and dehumidified. The fan supplies the air to the room through the supply air duct system. Room terminals (VAV, double-duct, reheat type) containing air dampers and/or heating coils are often used to permit individual temperature control of several rooms served by the same duct system. In winter, one must sometimes preheat the outside air, or mixture of outside and return air, to prevent damage to the chilled water coil by freezing. The heating coil (steam, electric, or hot water type) also provides additional heat to maintain the room temperature.

If humidification is indicated, a humidifier or water spray on the cooling coil surface evaporates water into the air stream.

To understand energy consumption in an HVAC system, consider the three elements that jointly determine energy requirements. These are:

Building thermal loads.
HVAC system response.
HVAC equipment performance.

Building thermal loads

An air-conditioning system maintains temperature and humidity within a building by adding or removing heat and moisture to balance the cooling and heating requirements or loads of the various spaces. These loads are of two types, sensible and latent. **Sensible heat** results in a change of temperature and can thus be sensed by a thermostat. **Latent heat** is heat in the form of vaporized moisture. It cannot be measured by its temperature, hence the term latent, which means hidden. Latent heat results in a change of moisture content and relative humidity. The sum of sensible and latent heat loads is called total heat.

An **interior room** in a multi-story building is usually surrounded by air-conditioned spaces on all sides, as well as above and below. Cooling loads come entirely from within the room and are therefore called internal loads. These consist of:

Lights: The sensible heat gain in BTU/hr is equal to 3.41 × the total lamp + ballast input in watts. No latent heat gain.

People: About 450 BTU/hr total heat gain per person, typically half of it sensible, the other half latent. This varies with activity and room temperature.

Equipment: Usually electrically operated, resulting in a 3.41 BTU/hr of sensible heat gain per watt. If moisture is evaporated or equipment is gas-fired, latent load will result.

Internal loads are not affected by outside temperature. They require cooling both summer and winter.

An **exterior room** has both internal and skin loads. Skin loads consist of solar gain and heat transmission through the glass, walls, roof and floors:

Solar: Sensible heat gain equal to the solar energy striking the glass reduced by a shade factor to make up for the effectiveness of the particular type of glass and shading devices.

Transmission: Sensible heat gain or loss equal to the area × "U" (heat transmission factor) × temperature difference between inside and outside temperatures. In calculating summer heat gain through outside walls and roofs, it is impossible in practice to separate the effects of transmission, solar gain, and heat storage. Instead one uses equivalents rather than actual temperature difference.

Most HVAC systems bring in **outside air for ventilation.** In addition, most buildings are subject to **infiltration of outside air** through cracks around windows and doors, and through wall cracks and joints.

Outside air: Sensible heat gain or loss = 1.09 × CFM × temperature difference between outside and inside. Latent heat gain or loss = 0.68 × CFM × difference in moisture content (grains/# air) between outside and inside.

Sensible load is the algebraic sum of all the above sensible items, and latent load is the sum of latent heat components. If the sensible load is positive (+), the system must provide sensible cooling; if negative (−), sensible heating. If the latent load is positive, dehumidification is required; if negative, humidification.

Loads during the cooling season usually require both sensible and latent cooling. If room temperature and humidity condition are to stay constant, the HVAC system must exactly balance sensible and latent loads. The ratio of sensible-to-latent cooling capacity of the system is determined by the coil surface temperature. A colder coil surface results in greater moisture condensation and a greater percentage of latent capacity. The load condition of the coil surface temperature is also related to the system's air flow. If the supply air is colder, less air is required to handle the sensible cooling load.

Typical HVAC System

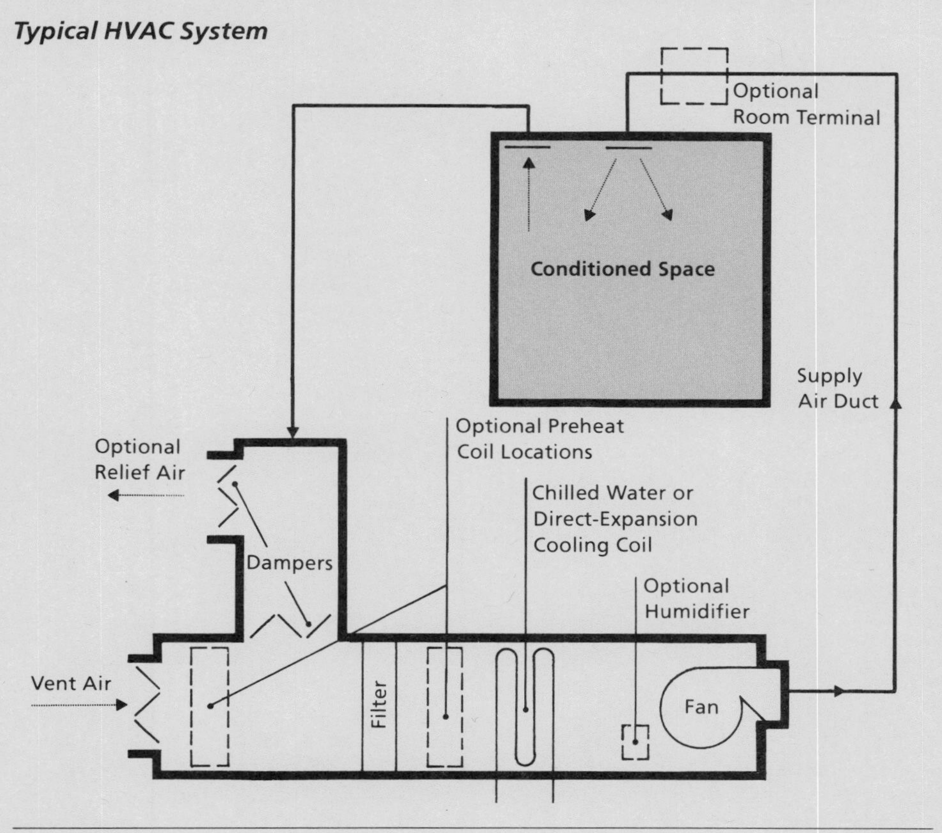

Figure 1-20

Part-load conditions

Actual loads at any one moment are seldom at their maximum values. Outside temperature, humidity and solar loads vary greatly, as does the wattage of lights and number of people. These factors, plus the tendency for equipment to be oversized, lead to a wide range of operating conditions and a low average load. The average cooling load in an exterior office may be 30% of equipment design capacity. An interior office load may average 65%.

Load and system capacity balance

Under most operating conditions, the HVAC system does not have the exact sensible and latent heat capacity to balance the calculated sensible and latent loads. If the sensible to latent ratios of the system load fail to equal the equipment capacity, the system cannot maintain the room temperature and humidity for which the loads were calculated, and room conditions will "drift." Drift will change both the load and the system capacity and continue until an exact balance is reached.

The latent cooling load will drop as room humidity goes up, due to the reduced difference in moisture content between ventilation air and room air. The equipment latent cooling capacity works the other way, with a rise in capacity as room humidity increases. Assuming that room temperature is being controlled by thermostat, it is humidity that varies. It will drift until a balance point between load and capacity is is reached, as shown in **figure 1-21.**

There can be no load without a system capacity to handle it. If a latent load is based on a 50% RH room condition and the chilled water temperature is raised from 45°F to 55°F to save energy, the coil surface temperature will probably also rise, resulting in less latent cooling capacity. The system will likely no longer be able to maintain the 50% design RH, so the room RH will rise until a new balance point is reached. Unless one takes the time to find the actual balance point – probably 70% RH – then the load will be overestimated. In this case, raising the chilled water temperature would save energy by reducing the latent load and by increasing the refrigeration machine's efficiency, but then room humidity would be unacceptably high.

Actual Room Humidity Determination

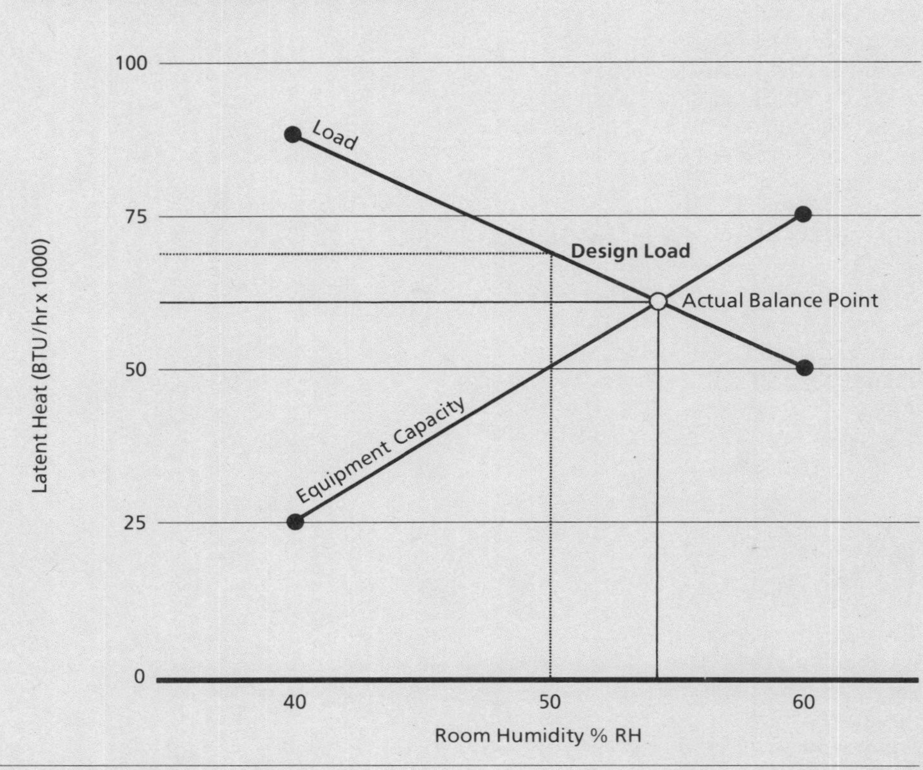

Figure 1-21

Most HVAC systems have some type of automatic **control,** usually a room thermostat, to reduce sensible cooling capacity until it equals sensible load at that moment. In most cases, this same thermostat also controls the heating coil. If the system serves only a single zone, cooling and heating capacity are usually controlled in sequence. Sequenced control occurs when a drop in room temperature turns off cooling completely before turning on any heating. This is the ideal method.

Most central type HVAC systems provide heating and cooling to a number of independent control zones. (In a high quality building, each room can be a separate zone with its own thermostat). One basic difference between these systems is the way they are arranged to provide this ability to serve many zones with independent temperature control from a common system. In many cases, they use some form of reheat to provide room temperature or humidity control. Reheat is heat added in excess of true room heating requirement at that time. It has the effect of increasing both the cooling and heating energy consumption by the amount of reheat, and should be avoided.

Reheat: sensible heat gain = 1.09 × CFM × temperature difference resulting from reheat.

The term **"system response"** refers to the reaction (or response) of a HVAC system to a building thermal load. It depends on the particular way the system is arranged and the building temperatures controlled. The first part of the system response process is to adjust building thermal loads for the inherent efficiency of the system arrangement and control, thereby determining the loads on the heating or cooling equipment. The second part of the system response is to determine air and water flow requirements of all fans and pumps that are parts of the system.

As an example of the first part of the system response process, let us assume that the outside temperature is 50° and the thermal load is 7,500 BTUH cooling in the area served by one zone of a conventional constant volume multi-zone type system. Analysis of the multi-zone unit control system shows that both cooling and heating energy is required to control the temperatures in the various zones. The zone in question has a small cooling load, but cannot provide the required cooling capacity without using both cooling and heating energy. The system response calculations would determine that 9,500 BTUH of cooling energy from the refrigeration machine and 2,000 BTUH of heating energy from the heating boiler is needed in order to provide the required 7,500 BTUH of cooling capacity to balance the zone thermal load.

A detailed discussion of system response and HVAC system control types is included in Appendix A.

Equipment performance

The equipment in most HVAC systems is larger than necessary. A 5% or 10% safety factor is usually added to the design cooling load before selecting equipment and, in most cases, individual items in the load estimate are conservatively estimated. As a result, cooling equipment may be 10 to 20% **oversized.** Heating equipment tends to be even more oversized, because a morning pickup allowance of 10% to 25% is often added to the design heating load. That is to provide extra capacity for warming the building quickly in case the heating system is turned down or off at night.

Consider the performance of each item of equipment at both maximum design conditions (for demand charges) and average part-load conditions (for energy consumption). Small compressors, fans and pumps are usually intended to be cycled on and off to reduce their capacity under normal part-load operation. The larger equipment is mostly designed for continuous operation. It is provided with a means of reducing capacity and energy consumption while in operation. The energy planner should not resort to cycling this type of equipment as a means of reducing energy demand or consumption without carefully evaluating the effect of shock and increased wear on equipment drives and starting equipment, due to frequent starting.

Basic Refrigeration Cycle

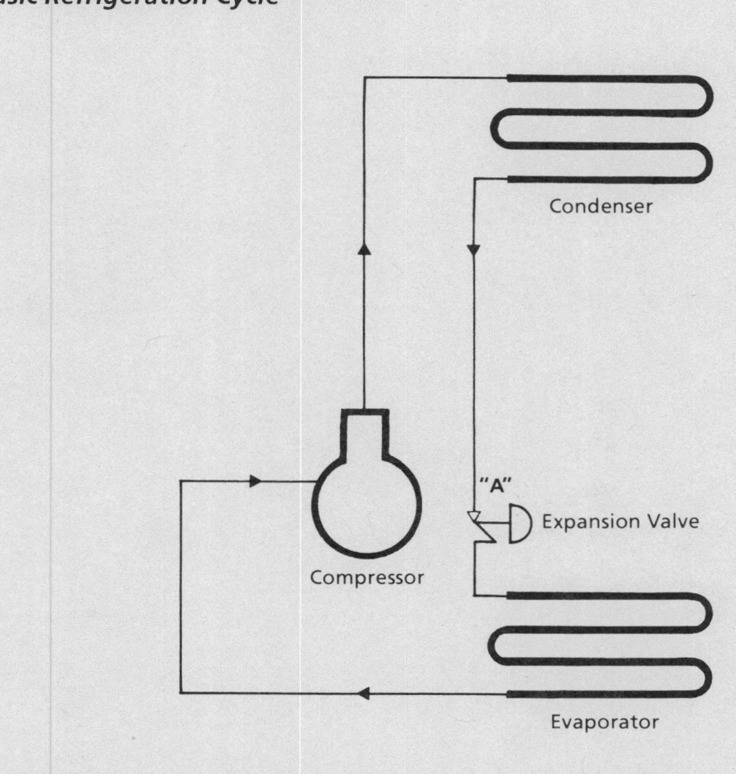

Figure 1-22

Refrigeration is usually needed to obtain the low cooling coil surface temperatures (50°F to 55°F) required for adequate cooling and dehumidification.

Most refrigeration machines are the mechanical compression type and use the **basic refrigeration cycle** shown in **figure 1-22.**

Refrigerant fluid is circulated by compressor and changes from liquid to gas at various points in the cycle. It becomes a liquid when its actual temperature drops below the saturation temperature (boiling point) at that pressure, and changes back to a gas as its actual temperature rises above the saturation temperature at that pressure.

Starting at point A, the refrigerant is a liquid at high pressure. The refrigerant flows through the expansion valve which reduces its boiling point (saturation temperature) to the temperature desired in the evaporator, which is sometimes the cooling coil in the air handling equipment. Air passing outside the evaporator coil is cooled by transferring its heat to the refrigerant inside the evaporator coil, causing the refrigerant to evaporate (at approximately 40°F). The refrigerant changes from a liquid to a gas in the evaporator. The gas is drawn from the evaporator by the compressor, which increases its pressure and discharges it as a high pressure gas to the condenser. The condenser is sometimes cooled by outside air.

The refrigerant boiling point (saturation temperature) was raised by the increased pressure and is now above the outside air temperature. Thus heat flows from the refrigerant to the outside air passing over the condenser coil. As the refrigerant cools, its temperature falls below the boiling point at that high pressure and it condenses back to a liquid. The high pressure liquid refrigerant drains from the condenser and flows back to point A, to start over again.

Instead of cooling the supply air directly, the evaporator coil may cool water, which is then pumped by a chilled water pump to a chilled water coil in the air system. This is a chilled water system, which lets a central water chilling machine serve a number of different air systems, with less risk of expensive refrigerant leaks. A single large refrigeration machine tends to be cheaper, easier to maintain, or easier to find a place for, than a large number of compressors at different locations.

Instead of using outside air to cool the condenser directly, a **water-cooled condenser and cooling tower** may be used. A **condenser water pump** circulates water from the cooling tower to the condenser, where it is warmed by heat rejection from the condenser. Warm condenser water returns to the cooling tower and is sprayed into the outside air. Part of the condenser water evaporates, thus cooling the remainder, which then goes back to the condenser water pump to begin another cycle.

An advantage of the water cooled condenser is that water from a cooling tower is usually colder than the outside air, so the condenser refrigerant pressure (saturation or condensing temperature) is lower. That way the compressor does not have to work as hard. Motor input power is usually about 20% less at design load.

Compressors are either reciprocating, for capacities up to 200 tons, or centrifugal, for capacities up to 10,000 tons. Screw type compressors are rarer, being available from about 100 to 500 tons. Choice of compressor determines the refrigerant used.

Full load performance of refrigeration equipment may be expressed as:

$$KW = Tons \times lift \div efficiency$$

In this case, **efficiency** is the product of three separate effects: the thermodynamic cycle efficiency of the particular refrigerant at the operating temperature, compressor efficiency, and motor efficiency.

The **tons** (1 ton = 12,000 BTU/hr) is the building cooling load. **Lift** is the difference between refrigerant condensing temperature and suction temperature, and depends on the type of system, as shown in **figure 1-23.** It also is a function of the relative amount of heat exchange surface in the refrigerant evaporator and condenser and the exact system design temperatures.

Compressor efficiency is determined by its design (piston speed, valve type and size, refrigerant gas passage geometry, impeller size, shape and speed, etc.). Motor efficiencies vary, and motor location (open or hermetic type) has a

Compressor Lift Requirements

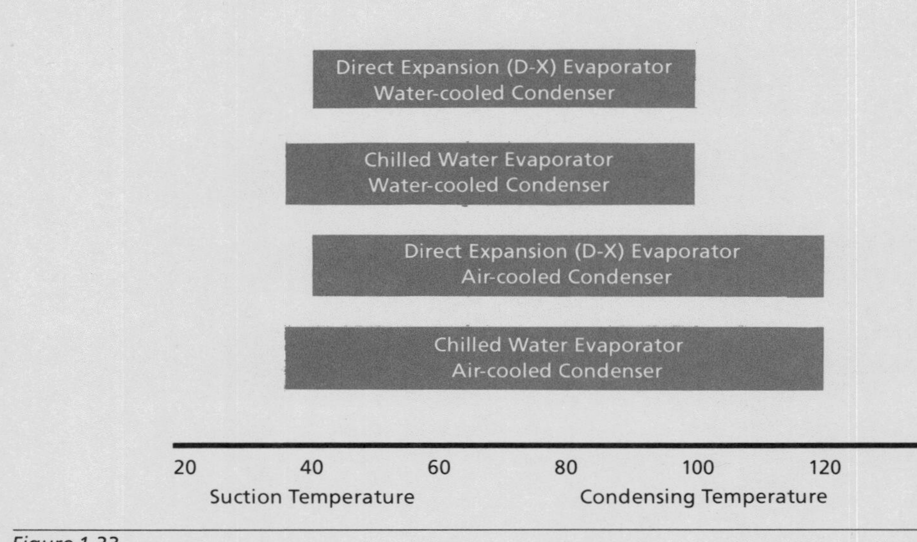

Figure 1-23

Typical Centrifugal Refrigeration Machine Performance

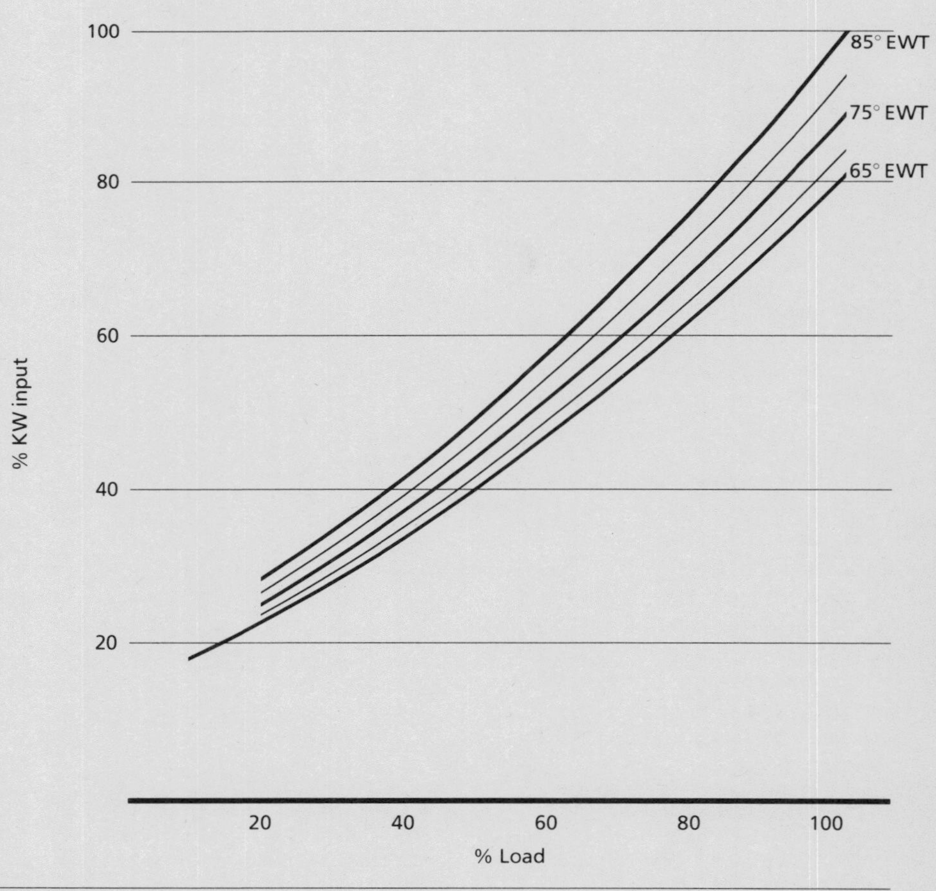

Figure 1-24

slight effect. In a hermetic compressor, the motor is sealed inside the refrigerant atmosphere. Motor inefficiency heat (8%) goes into the refrigerant instead of the equipment room, adding about 1.8% to the compressor power input.

Compressor part-load performance depends mainly on the type of compressor. Small reciprocating compressors are usually cycled for control, so the only performance adjustment is the effect of a lower condensing temperature and a possibly higher or lower suction temperature. Reciprocating compressors of over 15 tons capacity may come with either suction unloaders or discharge pressure bypass valves to obtain one or more steps of reduced capacity. The KW per ton at these steps is much higher than at full capacity, but this may be partially offset by improved heat exchanger performance at part load.

Centrifugal type compressors use inlet guide vanes to provide a continuously variable modulating capacity. **Figure 1-24** shows the **performance of a typical centrifugal machine.** Above 60°F outside temperature (40% load on a typical building), the power input per ton is less than at full load, but below this point it rises to a substantially higher level.

Absorption type refrigeration machines use steam or hot water to produce chilled water. The machines run with continuously variable modulating capacity and reduced energy input, but electrically driven auxiliary pumps on the machine work continuously at full load. They are used in fewer than 5% of commercial buildings. The *ASHRAE Data Book* or manufacturers' catalogs offer more information about both compression and absorption refrigeration equipment.

A **heat pump** is a refrigeration machine used for heating. Normally, a refrigeration machine provides cooling at the evaporator. A typical air-cooled electric drive machine requires about 1 KW motor power input to provide 1 ton (12,000 BTU per hour) of cooling effect. Heat rejection at the condenser is the sum of the evaporator cooling effect plus the motor input power. Heat rejection for each ton is:

Condenser heat rejection
= Refrigeration effect + motor input power
= 12,000 BTU/hr + 1 KW × 3410 BTU/KW
= 15,410 BTU/hr

Thus, there is 15,410 BTU/hr of heat with 1 KW power input. Using the 1 KW in an electric resistance heater would yield only 3,410 BTU/hr. It takes far less power to move heat with a refrigeration machine than to use the electricity in a direct resistance heater.

Most heat pumps are air-to-air. This means they use an air-cooled outdoor coil and an air-cooled indoor coil. During the cooling cycle, the indoor coil takes heat from the building HVAC system and rejects it through the outside coil. During the heating cycle, the outdoor coil cools the outside air (from 50°F to 40°F, from 20°F to 10°F, etc.) and rejects this heat, plus the motor heat, to the indoor coil at about 100°F to heat the building.

Heat pumps are better than electric resistance heat from a heating energy cost standpoint. Compressor wear is far greater, however, because it must operate both in heating and cooling seasons, and because the defrost cycles and heating season operating conditions are severe compared to ordinary cooling duty. Owners might expect shorter life and higher maintenance than from a product of similar quality used for cooling only.

Fans are used whenever air must be moved or circulated. Supply fans, return air fans, exhaust fans, cooling tower fans and air-cooled condenser fans are the most common applications.

The performance of all types of fans may be calculated from the formula:

$$\text{BHP} = \frac{0.000157 \times \text{CFM} \times \text{static pressure}}{\text{Fan efficiency}}$$

The CFM (cubic feet of air per minute) and static pressure (expressed in inches H_2O) are determined by the HVAC system requirements. Fan efficiency depends on fan design and the proximity of the actual operating point to the peak efficiency point.

Actual operating conditions of a fan may differ widely from design operating conditions. At any given fan RPM, a fan can deliver a wide range of air quantities. Fan inlet and outlet duct conditions will prevent the fan from operating exactly on its rated performance curve. Furthermore, the resistance of most duct systems tends to be less than anticipated because of safety factors and conservative estimating procedures.

With less actual system resistance than estimated, air flow can be reduced to the design value by closing a balancing damper, thereby adding a resistance equal to the difference between the design and actual system resistances at the design air flow. A better approach for conserving energy is to reduce the fan RPM.

Most commercial HVAC systems are arranged for continuous **fan operation,** so as to provide air motion and ventilation at all times. Constant-volume systems run continuously at full CFM; variable-air-volume (VAV) systems operate continuously, but at reduced CFM and possibly reduced static pressure, as directed by the system load. Small fans may be cycled to reduce energy consumption.

Obtaining variable air flow with constant fan speed depends on the **type of fan.** Airfoil or backward-inclined type fans can use inlet guide vanes to reduce the fan power, roughly in proportion to the CFM reduction. Forward-curved fans use discharge dampers for similar performance. Axial flow fans with variable pitch blade control will provide even better part-load performance.

Variable speed control has the lowest energy consumption for both airfoil and forward-curve centrifugal fans, because fan power varies approximately as the cube of fan speed, while air quantity varies as the first power of speed. Overlooking the effect of automatic dampers in the systems and the change in fan motor and drive efficiencies, 67% CFM could be obtained with 30% power input.

Two-speed fans save power through speed control, but do not provide the modulation needed by a typical VAV-type system.

Most large HVAC systems require circulation of condenser water, chilled water or hot water through piping systems. Centrifugal type water pumps are almost always used for this purpose, with operating performance calculated like that of fans:

$$\text{BHP} = \frac{\text{GPM} \times \text{head}}{3960 \times \text{pump efficiency}}$$

The GPM (gallons per minute) and head (in feet H_2O) are determined by the HVAC system requirements. Typical pump efficiencies range from 60% to 80%.

Pumps are also similar to fans in that the actual operating condition always occurs at the intersection of the pump performance curve and the actual piping system resistance curve. If actual system resistance is lower than estimated, or if required water flow is less than the design value, water flow can be reduced by closing a valve. This raises the piping system pressure drop, but slightly reduces pump power input.

Most systems are arranged for continuous pump operation. Savings may be made, however, by use of **variable water flow,** obtained by use of any combination of multiple pumps, two-speed pumps and throttling valves. The method used must provide for adequate water flow to all items of equipment that remain in operation at this part load condition. Pump characteristic curves of GPM vs. head vary widely, so the specific pump curve must be used in evaluating a variable flow water system.

Electric heaters and boilers are simple devices to estimate. Heat output is always 3.41 BTU per watt input, but boiler and water piping insulation heat losses should be considered for the hydronic systems.

Electric heaters and boilers can be single or multiple step type. Electronic continuously variable modulating controls also are available.

Fossil fuel boilers are usually fueled by natural gas, fuel oil, or coal. Still, almost any combustible material may be used. Auxiliary motors are required on oil and coal burners, and induced or forced draft blowers are usually required to remove flue gas. Make-up air is required for combustion. This may affect the heating requirements of boiler rooms or adjacent spaces. Burners may be either cycling "on-off" type; cycling "hi-lo-off" type, or modulating. Burner efficiency should be obtained from the boiler manufacturer; 75% is typical for oil and gas types, not counting boiler and pipe insulation losses, which can get to be a large part of total input at low speeds.

Humidification equipment adds moisture to the air stream. Heat is needed to evaporate this water (approximately 1,000 BTU per pound) and must be estimated whenever humidification takes place.

Steam grid or steam spray humidifiers inject live steam directly into the air. Pan type humidifiers consist of a shallow pan filled with water. A steam or electric heater is immersed in the water and energized to raise air humidity.

Evaporative spray type humidifiers obtain heat for evaporating water by cooling the air stream. The total heat content of air entering and leaving the humidifier is constant. Air entering the humidifier is relatively warm and dry; when it leaves the humidifier it is cooler, but more moist. In some systems, entering air or water must be heated to keep leaving air from becoming too cold. This heat must be calculated as part of the system's energy consumption.

In other systems, air entering the sprays is a mixture of cold outside air and warm return air. If the proportion of outside air can be reduced, the required mix may sometimes be obtained by taking more return air and less outside air. In this case, the energy to evaporate the water is free, as it comes from the increased amount of return air which would otherwise have been wasted to the outside as relief air.

Alternate energy sources

Economizer cycles, heat recovery, solar energy or waste heat should be seen as means to reducing required cooling and heating.

Economizer, or "free-cooling" cycles, provide cooling to areas of a building which require cooling when it is cold outside. Interior areas and rooms with south-glass exposure are typical spaces with year-round cooling loads. The most common method is the "outside air economizer cycle," in which the outside air intake is enlarged to permit use of up to 100% outside air whenever it is colder than the return air.

Waste heat is heat already available at a suitable temperature, such as building air at 70°F which is to be exhausted to the outside. If there is a need to preheat ventilation air coming in at 10°F, a coil may be installed in each air stream. Brine circulated through the coils will cool the exhaust air and transfer this heat to the ventilation air. The only utility energy required is the electricity to run the brine pump. Direct air-to-air heat exchangers are used for the same purpose; they also cool ventilation air in hot weather. Rejected heat from computer room or cold storage refrigeration equipment that runs summer and winter is often available, as is waste heat from industrial processes.

Heat recovery can refer to the direct use of waste heat, or to the use of a refrigeration machine to raise the temperature of a waste heat source to a more suitable level.

Any heat recovery system is better if both ends are put to good use. One should try to find a way to cool a space or item that has to be cooled anyway and use the heat for an item that needs heat.

Solar energy can be used for heating to provide heat energy to power a heat-operated refrigeration system, such as an absorption type system. (The subject is treated extensively in the references.)

The potential of utility energy reduction is limited only by the owner's finances. The art of practical **heat recovery** calls for heeding these few fundamentals:

Temperature level. With heat recovery refrigeration machines energy input is reduced by keeping heat output temperatures low, but this is not always practical. The proportion of heat requirements that can be satisfied may be increased by raising heat output temperature, especially in case of water heating. Output water temperatures of 125°F are common and temperatures of 180°F are available.

Optimum capacity. Most heat requirements go up when outside temperature is low, and most heat sources go down with lower outside temperature, resulting in a balance point, as shown in **figure 1-25.** The **heat recovery system should be sized** for this balance point unless a storage system is provided so heat can be stored at one operating condition and used later. A refrigeration machine sized for maximum summer HVAC load will usually be so oversized as to be impractical for use as a heat recovery machine.

Unnecessary work. Useful recovered heat is ordinarily a small percentage of the heat rejected from a heat recovery machine, so the rest must be wasted. If there is an outside air economizer cycle, less energy is consumed by forgetting heat recovery. For example, operating a 300 ton centrifugal machine having a 30% minimum load requirement is useless if the heating load is only 25 tons.

Adding heat recovery can penalize the HVAC system's cooling season efficiency. For example, a centrifugal refrigeration compressor is normally selected for greatest efficiency at a summer operating condition of 95°F leaving condenser water. If the compressor is designed to operate as a heat recovery machine producing 105°F hot water at the 20% to 40% tonnage range typical of winter operation, the compressor efficiency during typical summer operation will be from 5% to 15% lower. The heightened summer demand charge and energy consumption may offset most of the savings in the heat recovery season operation.

Idle investment. If a savings in utility energy cost is expected to justify the higher initial cost of a heat recovery or solar system, it is unwise to have so large an investment sitting idle. A system to heat domestic hot water or provide reheat energy for a reheat type system can be used year round and will probably pay off faster than one which only provides heating for a few months in winter.

More complete information is presented in the ASHRAE Data Books (Ref.).

Heat Recovery Balance Point

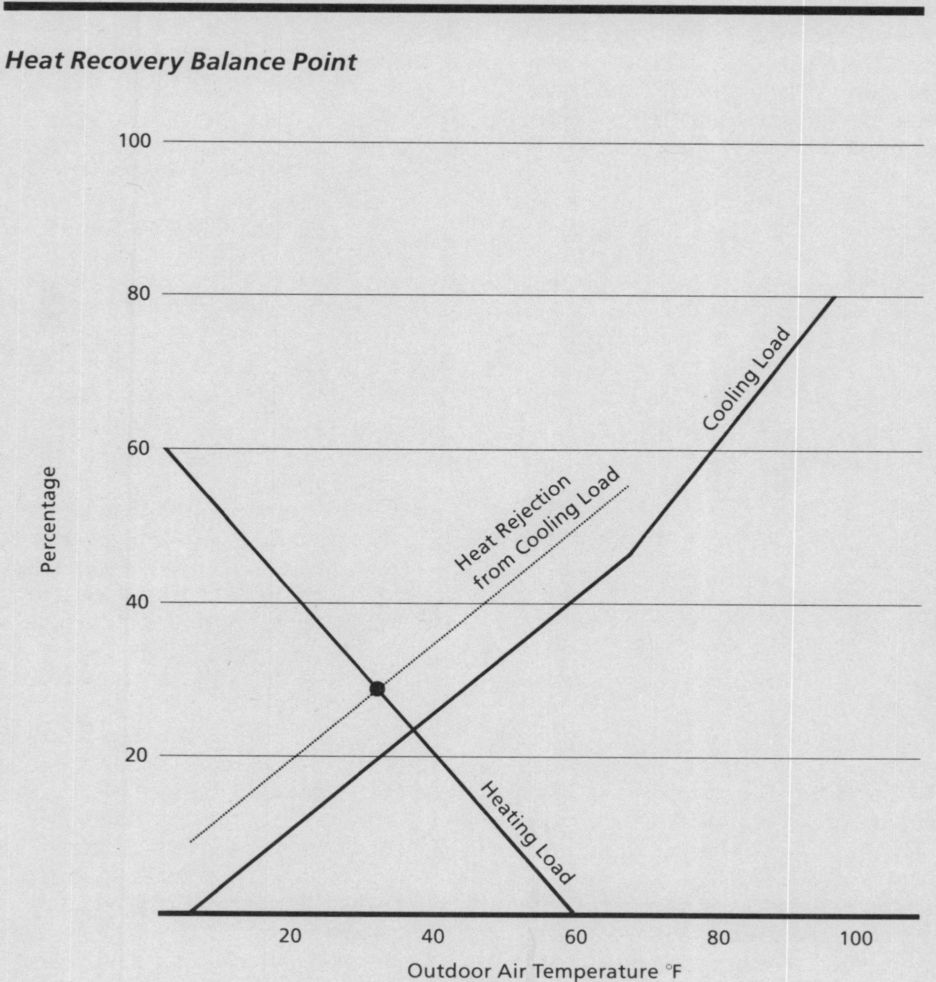

Figure 1-25

Energy estimating methods

The energy planner must estimate the energy consumption of many types of equipment.

Some equipment operates at the same energy consumption rate whenever it is in operation. The energy consumption is found by multiplying the consumption rate times hours of operation.

Other equipment operates at variable rates. At least four estimating methods are available:

- Equivalent full-load hours
- Degree days
- Hour-by-hour
- Outside temperature bins

Equivalent full-load hour method

The "equivalent full-load hour" method calls for judgment, job measurements or data from previous projects to estimate the equivalent number of full-load hours of operation per month for each item of equipment. These hours are multiplied by the hourly full-load rate of energy consumption.

This yields the required monthly energy consumption. The estimate is improved by using average load efficiency instead of full-load efficiency of the variable load equipment items. Still, accuracy of the results depends mainly on the user's judgment. This method is best for estimating consumption of equipment such as elevators and office equipment. In HVAC energy estimating, it is useful for making quick, rough preliminary estimates.

Degree-day method

The "degree-day" method is common in residential work, chiefly for heating season estimates. It establishes a base temperature, usually 65°F, below which the building begins to require heating. Heating requirements are calculated for a 24-hour design load heating period. The results are divided by the difference between the 65°F base temperature and the outside heating design temperature to obtain the BTU requirements per degree day. This value is multiplied by the number of monthly or annual "degree days" from weather data. These loads are then divided by the furnace efficiency, to estimate the energy consumed to meet the load.

This is an empirical method based on statistical samples of large numbers of buildings, with no assurance of accuracy when applied to a specific project.

Most HVAC systems are designed to handle extreme heating and cooling weather conditions which occur less than 5% of the time. The remaining 95% of the time they operate less efficiently at part loads, with the average load at about 30%. A system which is highly efficient at full-load is not necessarily efficient at reduced load. A proper HVAC energy estimate must, therefore, be "dynamic" so it can evaluate system performance over a full range of likely operating conditions. Dynamic analysis methods include the "hour-by-hour" and "bin" methods.

Hour-by-hour method

The true "hour-by-hour" method computes the instantaneous building load, residual stored loads, and resulting HVAC system performance separately for each of the year's 8,760 hours. It then adds them to obtain monthly and yearly consumption. Maximum demand (rate of energy usage) is found by identifying the hour with the highest usage. Building heat storage and temperature swing can also be simulated. Due to the amount of data handled, hour-by-hour analyses are often done by computer.

A modified hour-by-hour analysis may be done for one or more typical days of each month. The results for each day are multiplied by the number of assumed typical days to obtain the monthly energy consumption. This averaging process reduces the accuracy of the estimate by eliminating extremes of weather conditions, solar, and internal loads.

Outside temperature bin method

The "outside temperature bin" (a temperature bin is a range of outside temperatures, usually 5° increments) method is based on the principle that the load on an HVAC system is related to outside temperature. Like the "typical day per month, hour-by-hour" method, it uses an averaging process and is therefore less correct than the true full year "hour-by-hour" method. It does take into account internal loads, energy use created by patterns of operation and occupancy, and varying solar loads. The HVAC energy consumption at a few outside temperatures is calculated and the consumption at other temperatures and conditions is extrapolated. The bin method is illustrated in this book and includes four steps:

- Plotting the energy profile (a graph of energy consumption vs. outside temperature).
- Finding the number of hours in each outside temperature "bin." **(See figures SP-7A-B).**
- Determining energy consumption for each temperature bin.
- Totaling energy consumption for bins.

A separate peak demand estimate is required if utility demand charges are to be figured.

Computer programs

Computer programs are available using the hour-by-hour method, the typical day per month, hour-by-hour method, and the bin method. They vary widely in ability to provide accurate results for a particular situation. A program may provide an adequate solution for one situation and an inadequate solution for a different type of building or a different type of HVAC system. The inherent accuracy of a computer program depends upon:

- Analysis method, with the true hour-by-hour method having an advantage.
- Accuracy of load estimating. Many of the load estimating methods incorporated into computer programs were developed to conservatively predict maximum design load conditions, and are inaccurate in estimating part-load conditions.
- Accuracy of system response and equipment performance. Many programs are weak and inflexible in this area. There are at least a dozen types of variable air volume systems available, yet most computer programs are unable to distinguish one from the other.

Computers are able to manipulate a great number of weather conditions and perform a large number of error-free calculations. However, the program logic is frequently hidden or described in such a manner as to make it difficult for the user to understand, resulting in possible misapplication of the program.

In the case of energy analysis, "accuracy" is a relative term which is virtually impossible to measure. There are always differences between actual and estimated weather conditions, usage and occupancy, hours of building systems operations, control point settings, and building and system maintenance conditions which both create and explain differences between actual and estimated energy consumption. Annual energy estimates of the same building using the popular computer programs will vary as much as 30%. A manual estimate using the modified bin method described in this book will generally fall within this tolerance range, and therefore, can be considered equally "accurate."

The energy planner needs to be able to prepare analyses using both manual and computer methods. Simple preliminary analyses can usually be done quicker and with less expense using a manual method. A manual analysis is your only accurate alternative if you have an unusual situation or system which has not been adequately programmed for computer use. The computer is the best choice for presenting a final polished analysis, especially if many alternatives are involved.

Summary

This chapter was designed to provide the reader with a basic look at the concepts and concerns that enter into any energy planning effort, whether design of a new building is involved or the updating of an existing, energy-inefficient facility. Ensuing chapters take up the various phases of energy-conscious analysis and design in greater detail.

The Team: Roles & Responsibilities

Major responsibility for the energy planning project lies with the energy planner. The planner must know how to apply the architectural and engineering skills that will improve energy performance of a building, how to do a comprehensive analysis of the building's present performance, identify opportunities for improvement and evaluate these, and narrow down the list to those that should be carried out. Beyond this, the energy planner should be prepared to complete final designs and help the building owner with construction management and developing a continuing building operation and maintenance program.

The building owner is responsible for financing the project, selecting the energy planner, providing data, suggestions and access to the building, reviewing and approving progress, and coordinating staff and the building's users, operators and contractors throughout the project. It is up to the owner to see that the energy planner's recommendations are carried out and continued energy efficiency maintained.

Introduction

Every energy planning project is a team effort. The energy planning firm selected to perform the work will need cooperation from all who may own, use, operate and maintain the building, as well as from the contractors and suppliers who carry out the energy planner's recommendations.

The energy planner

An energy planning firm is a design firm with expertise in analyzing and improving the energy performance of buildings. This expertise is emerging as a multi-disciplinary specialty reflecting concern for both the architectural and engineering aspects of energy-conscious design. A close working relationship between the traditional design disciplines is essential. If successful, it will prod the building designers to explore for more imaginative ways of controlling a building's environment.

An energy planning firm may consist of one person with a wide range of skills, or a group of specialists who collectively know all those aspects of a building which shape its energy performance. The firm should also know about the use of natural daylighting and how to take advantage of orientation, landscaping, fenestration, and sun controls. Members should know how to design energy-efficient artificial lighting and how to integrate natural daylighting with artificial. Such energy-conscious design skills coupled with creative mechanical engineering can reduce the need for mechanical heating and cooling and still provide comfort while spending the least amount for energy. Other attributes include an understanding of life-cycle costing and the financial implications of energy-conscious design.

A building's energy performance is determined by a complex and unique set of variables. The most valuable road to improvement usually results from detailed analysis combined with creative problem solving. The more comprehensive and integrated the approach, the better the solution.

Traditional design firms develop energy planning expertise in order to:

- Produce buildings which are more energy-efficient, comfortable and attuned to their environment.
- Increase their scope of offered services.
- Satisfy required energy and operating costs of government agencies and private clients.
- Take advantage of related opportunities such as teaching, consulting, research, testing and product development.

Energy planners offer a comprehensive service of analysis, design and construction management to improve a building's energy performance. Since they sell services, not energy-related products, they are able to consider objectively all available approaches.

The owner

Even more than on conventional architectural projects, owners must take an active role as the project's client. Their first responsibility will be to select an energy planner. Initial discussions focus on the project's scope, budget, time schedule and approach. Often, project funds are limited and the owner is confused by a myriad of attractive advertised solutions. Product representatives leave mailing cards, each claiming his product will greatly lower annual operating costs.

The experienced energy planner can show the owner that many of the **building's environmental support systems are closely linked.** A change to one part may have drastic effects on another and, thence, on performance of the entire building. Modifications which save energy can increase operating costs, shorten the life of equipment, and create larger problems.

For instance, although cycling large fan motors on and off may reduce their energy use, doing so can wear them out prematurely if they are not designed for this manner of operation.

Another responsibility of the owner from the start is **financing.** A unique situation may arise since funds are being spent to reduce annual operating costs and are usually expected to be recovered within a few years, often as few as one or two. After the initial payback-period, savings will accumulate. Therefore, whereas an owner who hires an architect ends up with a new building, the owner who spends dollars to modify an existing building to save energy and money ends up with an "investment." His approach to the project therefore reflects a **"return on investment"**

The building's users, operators, energy suppliers, contractors, and product manufacturers are valuable members of the project team. They provide suggestions and information as they carry out the energy planner's recommendations, as well as beforehand.

Specific project responsibilities of the owner and energy planner are described in the contract along with a clear definition of the work to be done, schedule and compensation.

In preparing the proposal and subsequent contract, the energy planner should determine what the owner expects to achieve and what resources are available, and balance this with:

The annual savings likely to result from the project.

The level of investment the owner considers to be reasonable to achieve those savings.

The time and effort the owner is willing to devote to the project.

The owner's priorities.

frame of mind, tracking anticipated return compared with alternative investments.

In a leased building, financing arrangements for a redesign can vary from 100% owner financing to 100% tenant financing, or a combination of the two. In either case, a review of leases may be required. This process must begin early in the project or implementation may well be delayed. The owner should make sure tenants understand the project's goals and are kept up-to-date as to projected costs, savings and expected improvements. If new **leases** are required, negotiations should get under way while alternatives are still being evaluated. There is no point in time consuming analyses of alternatives unless these are acceptable to tenants.

Here are some common lease situations:

Present leases contain no building operating cost escalation clause. The owner is caught between spiralling costs and level income. Such owners can afford to absorb the costs of energy analysis since they alone will benefit financially from the improvements.

Present leases contain a building operating cost escalation clause. Tenants will benefit financially from improvement. A portion or all of the project costs may be pro-rated among them.

Present leases are based on the Consumer Price Index. This situation depends on the relationship between the CPI and the rate of increase in utility costs.

Both owner and tenants stand to gain from lease arrangements resulting from shared energy project costs because:

The owner's expenses are reduced.
Operating costs are reduced.
A more competitive lease can be marketed in the future.
Tenants are more likely to cooperate if they have an investment in the project.

By aiding in the project, tenants can expect:

Lower net rental cost as amortized over the term of the lease.
The building's environment should be improved.
When the lease expires, they may renegotiate for a lower price than in comparable buildings with unimproved energy performance.

Once the project is under way, **the owner's main responsibilities** can be stated as follows:

Provide the energy planner with the information (drawings; specifications, if available; utility records; operation and maintenance schedules).

Provide the energy planner with access to the building, records and key personnel, as required to inspect the building and its equipment, inventories, and to discuss building use and operation with users and operators.

Review progress. This is especially important as each major task begins or ends. Data and assumptions used for the energy audit and energy analysis should be reviewed: inaccurate information can make an entire analysis worthless. The energy planner is advised to request the owner for written confirmation at the start of key phases of the project.

Coordinate the building's users and operators. At some point, the owner may wish to establish instruction programs to explain changes to the building's users and operators and show how these can contribute most to the project.

Contribute suggestions. The experience of owner and staff in using and operating the building serves to single out which aspects of the building's performance need improvement.

Carry out the energy planner's recommendations. This is the owner's most important responsibility. Improvements will only be realized if recommendations are adhered to. If construction is involved, owners assume the role typical of a conventional renovation. They will need to arrange financing, approve designs and samples, coordinate users, staff and contractors and, during the work, inspect and approve the work, with the energy planner's advice.

Follow through with efficient operation and effective maintenance. The owner must see to it that the energy planner's recommendations are translated into a continuing program of routine procedures and periodic inspection, and that the building's users and operators become aware of their new responsibilities.

The owner's role is time consuming, especially for a large building or building complex, so it may be advisable to appoint an owner's project manager to take over day-to-day responsibilities. The energy planner helps define the owner's responsibilities during contract negotiations. This way the owner can estimate the time and effort required, organize staff, and delegate responsibilities before the project starts.

Responsibilities of owner and energy planner should be clearly defined in the owner/energy planner agreement. The energy planner will prepare a proposal outlining the project's scope of work, division of responsibilities, schedules and compensation. Once accepted by the owner it may be incorporated into the contract.

The building's users

The building's users can become very helpful if made aware that **the goal is to make the building not only more energy efficient but also more pleasant and comfortable.** In addition, their experience with the building may help identify opportunities for improvements, and their cooperation will be crucial for continued energy efficiency. Specific contributions include:

Assessments of the building's present performance.
Users can tell the energy planner how comfortable the building is and how well it serves its purpose.
Information concerning the building's use and operation.
Identification of areas of discomfort and other items that need improvement.
Suggestions involving design solutions and changes in operating policy.
Cooperation during testing and construction.
Continued sympathetic use of the building. An example of this is proper use of sun control devices, lighting and HVAC equipment controls. Changes in routine and/or changes in the internal environment may be necessary. For instance, reduced building hours or lighting redesign may call for special cooperation and support from users.

Surveys may help elicit user support as well as information. Questions to include may take up comfort, use patterns and suggestions for improvement. Surveys can be enhanced if kept brief and if followed by interviews with users.

The building's operators

A building's energy performance is no better than the quality of its operation and maintenance. Because of their experience with the building and its users, operations and maintenance personnel are the most direct source of information. They also will run the building after the work is done, so total cooperation of both operation and maintenance staff is vital. Specific contributions can include:

Information on the building's history, operation, maintenance and use.
Access to the building and its equipment.
Suggestions for improvements.
Cost estimates for modifications.
Assistance in installing and monitoring testing equipment.
Cooperation and coordination during construction.
Carrying out recommended modifications.
Conscientious operation and maintenance of the building.
Monitoring the building and serving as a source of "feedback."

Energy suppliers, contractors and product manufacturers

Energy suppliers or utilities represent a **local base of experience** as they deal with so many buildings. Contractors are acquainted with the features of the equipment they install. Product manufacturers or their sales representatives know their particular line of equipment and can spell out what their new products can do for the owner.

These groups can help the more detailed energy analysis along by contributing:

Required data. Data involving energy sources can only be provided by the energy suppliers. Manufacturers' representatives may be the only source for the part-load performance characteristics of HVAC equipment as needed for the analysis.
Suggestions for improvements.
News of recent or forthcoming developments in equipment and materials.
Cost estimates for modifications.
Installation of samples.
Cooperation during construction.
Proper testing and adjustment of completed work.
Operation and maintenance information.

Building officials

Building officials should be consulted as modifications are evaluated to see what impact **regulations** will have. In many projects, no building permit is needed as there are no changes in the building's structure or egress. Codes must, however, always be considered in two areas.

First, a reduction in circulated air quantities may require a variance from building officials. To guide code groups in this respect, the American Society of Heating, Refrigerating, and Air-Conditioning Engineers (ASHRAE), is developing air change requirement standards that take into account the need for energy conservation.

Second, new materials should be checked for acceptable flame spread ratings. At times, a new product or material (such as a ceiling insulating board) may not yet have code approval. In that case, it is the supplier's or contractor's job to obtain such approval. The process should get underway as soon as the new product is looked at seriously, so that other options can be developed if approval is not likely.

The owner/energy planner agreement

A major purpose of the energy audit is to determine if the building warrants a more detailed energy analysis. The contract will usually reflect this two-phase approach. Compensation for the audit may be based on a lump sum or guaranteed maximum, whereas the energy analysis will require a cost-based compensation approach such as described in ***Compensation Guidelines for Architectural and Engineering Services, A Management Guide to Cost-Based Compensation,*** Second Edition Revised, 1978, published by the A.I.A.

In many instances, owners know how much they are willing to spend and the energy planner's task is to see how much can be achieved within that amount and whether it pays to go beyond that amount should the owner seek added savings.

The scope of work will depend on:

The amount of potential savings. (The greater the savings, the more the owner is likely to invest. Except that it is easier to save the first 20% than the next 20%).
The number of buildings included in the project.
The complexity of building systems and operation.
The level of resources available for the project.

Determining the owner's expectations and resources

The energy planner should learn **why the owner wants to undertake an energy planning project.** The reasons will determine the approach to take and the type of contract required.

If the owner needs to prove compliance with an energy regulation, the project's scope will be determined by the compliance procedures spelled out in the code. If the code is prescriptive, a plan check review against the requirements for a building's materials, configuration and equipment may be all that is needed. If the building must conform to performance standards, which simply define the amount of energy a building should use, some form of energy analysis approved by the regulatory agency will be required. Energy planners should also explore their own liability with regard to clients' anticipated results.

If the owner is interested chiefly in **reducing annual operating costs,** then developing the agreement becomes less simple. This is because the energy planner will need not only to estimate time and effort required for an energy estimate, but also to choose the type of estimate to be used and the degree of detail. Will an energy audit be enough or will a detailed energy analysis be required? Will reports need to be prepared? If so, how many and how elaborate? Does the owner wish to improve comfort conditions or appearance as well as energy performance? Which is more important: reduced annual energy costs or improved comfort? What criteria will be used for evaluating possible modifications?

The answers will notably affect the labor required of the energy planner and, hence, the character of the agreement.

The energy planner should **identify the owner's resources** for the project. The extent of the owner's own time and effort, as well as that of the owner's staff, should be looked into, along with the level of funds set aside for the project. There is a lot of work the owner's own organization can do with just a little direction, such as time-consuming data gathering, monitoring, setting up test areas, technical assistance and economic analysis. The same staff may also be capable of managing and/or carrying out the recommended modifications. This possibility should be reflected in the agreement.

Before submitting a proposal, the energy planner can sometimes **size up the potential.** For instance, a prospective client, a 100,000 sq. ft. hospital, uses $6.00/sq. ft. per year in energy costs. If many similar hospitals spend about $4.00 per sq. ft., the planner can estimate that the first year's savings resulting from a comprehensive energy planning project may reduce costs to $3.00/sq. ft., or $300,000.

If the owner is willing to accept a three-year simple payback period, a project budget near $900,000 could be considered, including all energy analysis work, design and installation of modifications. On the other hand, if the hospital has its own laundry, and laundries typically cost $1.00/sq. ft. per year for energy, the energy planner may wish to revise his predictions downward, then come up with a rough estimate for the project's budget that he can compare with cost predictions for the work involved.

This will lead to a project approach in line with the owner's resources and hence a realistic proposal.

Developing an approach to the project

Energy planning is made up of three steps: 1) studying the present performance of the existing building or the existing design for a new building, 2) identifying opportunities for improvement, and 3) evaluating all opportunities. If the project's budget is too low for a two-phase audit and analysis, the energy planner may need to spell out in the initial contract, in some detail, exactly what will be studied.

A small project may be too expensive **to analyze in depth.** Consider a 10,000 sq. ft. office building with $1.00/sq. ft. per year electricity costs totaling $10,000 per year. Suppose the energy planner estimates that $4,000 is spent for lighting and the remaining $6,000 for heating, cooling, ventilation and other equipment. If the latter costs can be cut in half, $3,000 will be saved the first year. If the owner has specified a two-year payback, there will then be $6,000 to work with, including implementation costs.

If the estimate also included a 20% reduction in lighting energy, it will add an additional $1,600 to the project budget, totaling $7,600. This will allow the energy planner next to nothing for the performance study, meaning that only a few well chosen opportunities can be explored. The energy planner will stick to methods requiring few calculations and little redesign, relying instead on a study only of the building's lighting, its operations schedule, and the controls for its mechanical equipment. The service will consist essentially of general recommendations. On a similar but larger building with a project budget of $76,000, detailed redesign and specifications would be included.

If the aim is to minimize the owner's peak amount of cash flow at any one time, the contract should show that the project will **proceed in phases,** with savings from a previous phase paying for a later phase. The two feasibility phases described earlier – the audit and the analysis – can then be followed by opportunity identification and evaluation and preparation of construction and implementation documents. These in turn would be completed in one or more additional phases and priced out.

Phasing of the planning and analysis work also reduces confusion on the part of owners who often want to know costs and savings before knowing the extent of the problem.

Once the energy planner has determined the best approach for the work, he or she can **develop a proposal.** This proposal should include not only a listing of tasks and sub-tasks, along with an estimate of time and cost required, but also progress payments and the nature of deliverable "products" such as reports.

Because projects may variously resemble renovation work, tenant fit-up, interior design, new construction or maintenance, a standard contract form does not exist. However, the AIA Standard Form of Agreement Between Owner and Architect for Special Services (B727) provides a form into which to insert the contract.

Contractual complications can arise, especially if the owner wants results which the energy planner is unable to guarantee unless provided with authority to operate and maintain the building. Factors may be beyond the energy planner's control, such as annual weather variations, changes in operation and equipment failure. The energy planner should therefore spell out design assumptions and carefully consider responsibilities with respect to specific costs estimates and predictions of energy savings.

The owner should realize that energy estimating for buildings is a relatively new and imprecise field. Even with computer simulations it is difficult to obtain close agreement between different estimates for energy use in the same building. Thus, the contract's language must be carefully worded to protect the energy planner from unwarranted exposure to errors and omissions, while still providing the owner with practical, credible results.

Evaluating Present Building Performance

The building's present performance provides a basis for identifying and evaluating opportunities for improvement.

This performance analysis has been divided into two parts: the energy "audit" and the more complete energy "analysis." The analysis is a natural extension of the audit in any energy planning project.

The energy audit

The scope of work, costs to the owner and time required are low when compared to the energy analysis.

The audit consists of examining the building and its past utility bills. Energy performance is evaluated by comparing such facts as the building's annual energy costs per square foot or measured lighting levels to those of similar buildings.

The audit does not provide enough information to identify reasons for poor energy performance, since its scope is too short and superficial. Therefore, when the audit is used, recommendations should be made with care and only about those modi-

Introduction

To obtain a basis for improving a building's energy performance, many aspects of its performance have to be studied. These aspects include energy and economic performance, comfort conditions, how the building functions, its environmental impact, appearance and compliance with applicable building regulations.

The ***energy audit consists of a study of the building's past utility bills,*** *the most easily available performance information. In the* ***energy analysis, these utility bills are then compared with an energy cost estimate based on results of a simulation of the building's energy use.***

Opportunities for improving the building's performance are noted as they arise. Some will be obviously worthwhile; others need evaluation based on a simulation of the building's energy use as provided by the energy analysis.

Factors affecting energy use are interrelated, *and inaccuracies in one area can invalidate others. Comparison with the building's past performance, including past utility bills, records for similar buildings and the energy planner's own experience all help to verify estimates. In case of discrepancies, basic assumptions such as operation and use schedules should be looked at again. Then, if all seems in order, review calculations for computational errors. The greatest opportunity for improvement may come up when differences between the estimate and actual bills are accounted for or resolved.*

Once the energy planner is confident of the results of the audit or analysis, the owner will want to reconfirm assumptions and to learn the limits of accuracy for any projected costs or savings.

A report is usually required. It should include key findings as to the building's present performance, energy cost, demand and consumption; installed equipment capacity vs. actual requirements; and conclusions regarding non-energy related items such as comfort. Analysis methods and assumptions may be recorded. Photographs help explain problem areas. Graphics are useful in displaying the outcome of energy cost, peak demand and consumption studies.

If the report is to serve also as a proposal for the next phases, it should describe the work needed to carry out recommendations and methods to be used, and outline a schedule and budget.

Performing the energy audit

An accounting of the building's energy use and costs may point to major opportunities for improving a building's performance. Tools include information about operation and use, equipment and energy use. As a first step in any energy planning project, an audit provides an early understanding of how a building performs compared to other buildings. This helps set the best approach for the more comprehensive energy analysis.

Government agencies which require energy audits specify the procedure to follow. Contact them for details.

The audit consists of these steps:

- *Collecting data.*
- *Estimating energy performance.*
  - *Studying energy costs.*
  - *Studying installed equipment capacity.*
  - *Studying peak energy demand.*
  - *Studying average energy consumption.*

fications the energy planner knows will improve the building.

The audit's best use is to help the energy planner and owner decide if an energy analysis is worthwhile and to identify problem areas. A large part of the work is done in the field.

The energy analysis

The scope of work is detailed enough to allow opportunities to be accurately identified, projections of costs and benefits to be made and methods of carrying out the work to be determined.

The energy analysis calls for analysis and simulation of the building's operation using methods such as the manual, modified bin method.

Both the energy audit and the energy analysis involve the following tasks:

Data on the building's energy performance, user comfort and function are collected, organized and verified. Installed equipment capacity is compared to actual requirements. Peak energy demand, average energy consumption and energy costs are studied.

Data collection focuses on a walk-through study, including interviews with members of the building's operation and maintenance staff, management, possibly the utility companies, as well as the owner and his staff. Include all who are to work on the project on the first walk-through, so they will develop a feel for the building and its performance.

During the walk-through, the building and its energy consuming equipment should be surveyed to see how it is operated and maintained. The field survey team should also note how the building and its various spaces are being used and how they are lighted, and note every opportunity for improving energy performance. Some will be evident, such as repairing building damage, shutting down equipment in unused areas, or reducing excessive lighting. These points will be confirmed by the energy performance estimate and may uncover added opportunities.

The audit may enable the energy planner to uncover specific opportunities which should be explored right away. It also will give a hint as to the resulting impact on the building's energy performance. The energy audit may well uncover other potential modifications which will require a more detailed study.

Here are the **key steps in an energy audit.**

Collecting data for the energy audit

Collection of data can become expensive and time consuming if not clearly directed to answering questions at a level of detail commensurate with the level of investigation.

Since we have defined two levels of investigation in this book, the Audit and the Analysis, the data discussion will also have two levels of detail. As one identifies specific opportunity areas, the data required may become more extensive.

The extent and type of data collected is further influenced **by time allotted** in the audit scope of work, the ease with which it can be gathered and the **sensitivity of the results** as applied to the data being sought. For example, there is no point in gathering extensive U-values if the building shell has little influence on energy use.

The scope of data collection also is shaped by the **degree of success expected to result** from the study. No great amount of data is needed to see that a building is already operating very well.

Data for an audit includes:

Building

- Floor area and locations and types of uses for air-conditioned, heated space and non-conditioned spaces.
- Glass areas, including skylights, their orientation and shading (note type of glass and operation of windows).
- Wall and roof surface areas; thermal resistance; and weight (in case of massive buildings).
- Building orientation, configuration and solar access.
- Drawings, equipment schedules and specifications for mechanical system design and distribution layout, including control diagrams for operation modes.
- Electrical drawings and reflected ceiling plans for lighting layout, fixture type and quantity.

A walk-through or field survey of the building will let one note opportunities for improvement such as:

- Adequacy of the building envelope as indicated by signs of moisture penetration, deteriorated insulation, poorly sealed openings or openings that will not seal.
- Problem of comfort including poor temperature control, noisy equipment and poor lighting.
- Level of maintenance as indicated by loose fan belts, clogged filters and leaks.
- Discrepancies between contract documents and building: were the specified types of systems and equipment installed?

Environment

Weather information can be obtained from:

The local office of the National Weather Service, the Environmental Data Service of the U.S. Department of Commerce National Oceanic and Atmospheric Administration (NOAA). Monthly high, average and low temperatures should be collected to correspond to the years of the building's utility records. Monthly heating and cooling degree days may also help comparison with similar buildings in other areas.

Equipment

Plans and specifications should note major pieces of equipment and their installed capacity. Building operators often keep detailed records and catalog "cuts" of equipment for preventive maintenance. Where drawings are the only source, however, **equipment nameplates** may be checked to verify the drawings.

Rated performance and rated energy input should both be obtained for major equipment. Sometimes **field readings** must be taken to verify actual conditions in use. The walk-through will give the energy planner a chance to:

- Confirm information obtained from the drawings and fill in any gaps.
- Check the equipment's condition and look for signs of disrepair.

Figure 3-1 shows energy consuming equipment in a typical building. In computer centers, industrial buildings and restaurants, process and production equipment will consume larger portions of energy entering the building than in offices. Such a category on the chart may then become the most significant.

Operating methods

The energy planner also must know **when and how** each major piece of equipment is operated and maintained. Since operating schedules cost least to change and may produce significant savings, at least a rough operating schedule for each piece should be made to see how closely **operation of the equipment matches use** of the building (one example is shown in the sample problem in **fig. SP-11).**

Gathered information should include:

- Building operating hours (weekend, weekday).
- System and equipment control schedules.
- Control set points, such as discharge air temperature, indoor temperatures, chilled water temperatures and freeze protection temperatures.
- Rooms requiring special temperature control, such as computer facilities.
- Preventive maintenance schedule and records of recently replaced equipment.
- Quantities of exhaust, outside ventilation and air movement.

This data provides a good picture of the condition of the equipment and operation.

Past performance

Obtain from the owner **monthly utility bills** for the past two to five years, including:

- Energy peak demand and billing demand (if different from monthly peak demand).
- Energy consumption.
- Energy costs (demand charges should be separated from consumption and fuel adjustment if possible).

This data can come from the building owner, or from utility companies at the request of the owner. Monthly figures

Energy Consuming Equipment

Purchased Energy Consumers

- **HVAC**
  - *Heating*
    - Primary
      - Furnaces
      - Boilers
    - Distribution
      - Fans
      - Pumps
  - *Cooling*
    - Primary
      - Compressors
      - Cooling Tower Fans
      - Cooling Tower Pumps
    - Distribution
      - Fans
      - Pumps
  - *Ventilating*
    - Fans
- **Lighting***
  - Lamps
  - Ballasts
  - Dimmers
- **Vertical Transportation**
  - Elevators
  - Escalators
- **Domestic Hot Water**
  - Primary
    - Heaters
  - Distribution
    - Pumps
- **Process Loads***
  - House (Domestic Water) Pumps
  - Production Equipment
  - Data Processing
  - Miscellaneous Equipment

*Separate inventories should be taken of the equipment and lighting in (1) conditioned spaces, (2) unconditioned spaces and (3) outdoor areas so that proper allowance can be made for heat gain to HVAC loads.

Figure 3-1

should be correlated with the actual month of use, not with the month in which the meter was read or payment was due.

Utility companies keep and **compare records of energy costs of similar buildings** and their rate structures. They also can provide information about the basis of a particular building's rate structure. Facts about the building's installed capacity as compared with similar buildings may point up oversized equipment, irregular operation patterns, and/or unusual energy consumption that should be looked into further.

Energy sources

Identify the building's energy sources and obtain a copy of applicable current rate structures, along with projections of **future costs** and rate structures. (Utility rate structures are discussed in more detail later in this chapter.)

Comfort and other considerations

How the building is performing with respect to the other areas of concern may need to be explored. For example:

- Records and conversations with occupants may identify serious comfort complaints.
- Rental rates and vacancy statistics may be a sign of problems in renting or using the building.

Studying energy performance

Compare actual energy performance, as shown in utility records, with performance standards, or other similar buildings. Also, rough **calculations of theoretical energy use** may point to opportunities for improvement.

Here are areas of concern:

- Energy consumption per square foot.
- Peak demand per square foot.
- Installed HVAC capacity per square foot.
- The "load factor," or energy consumption divided by the product of peak demand and total hours.
- Energy costs per square foot.
- Energy use per unit of production (commonly used in manufacturing processes).

Performance may also be compared from one month to the next or one year to another. This will show when energy use is most intensive and, to some extent, the characteristics of why and how this energy is used.

Direct cause/effect is rarely evident simply from reading and plotting the bills. The step will, however, serve to pinpoint glaring problem areas worth pursuing in detail, and to hint at how the building compares with others.

Studying energy costs

The purpose of this step is to **identify costs that are out of line.** In addition, relationships may appear caused by fluctuations due to changes in weather, rate schedules, occupancy, type of controls and building operation.

The procedure is to **plot and compare two to five years of monthly energy costs** for each energy source, including peak demand charges and fuel adjustment charges. The energy planner may then be able to form an idea as to possible improvements, namely, those which affect the building's response to weather (adding insulation and double-glazing), those dealing with the building controls (time clocks, equipment set points), and those affecting building occupants (scheduling of building use).

Comparing the data will disclose whether heating or cooling is the critical concern and, if so, when. For example, the relation between energy cost fluctuations and weather changes will be direct only if the building is operated the same hours each month, if utility costs are constant all year, and if there are no reheat coils in the cooling system.

The energy planner should avoid making recommendations before carrying the investigation far enough to draw conclusions.

Figures 3-2 and 3-3 show monthly energy costs for a typical 145,000 sq. ft. gas-heated, electrically cooled office building, operated according to the same schedule all year and without any form of reheat.

From the monthly graphs on energy costs one can draw several **conclusions.**

Monthly energy costs (figure 3-2):
Monthly costs range from about $9,000 to $14,000 per month. Least costs occur in spring and fall when outside air can be directly used in lieu of mechanically refrigerated air, and heat losses are low.
Electric costs are several times higher than natural gas costs. Electric costs are thus the first concern.

Monthly electric costs (figure 3-3):
Electric costs do not vary much from month to month (20%), and are highest in summer.
Even electric demand costs are several times heating costs. Demand costs are flat for much of the year, except for a 10% escalation during August. Perhaps the demand charge rate went up, or actual demand.
January electric consumption dropped sharply from December, even though January tends to be colder in this location. This was perhaps due to shortened operating hours in January or extended hours in December.

Monthly gas costs (figure 3-2):
Gas costs are far higher in winter. Gas use is linked most directly to weather and the graph shows this.
Perhaps a third of natural gas costs are for use in the summer. Cause may be a service charge or an inefficient domestic hot water system. More detailed information can be developed by comparing this one year of data to previous years, or to other buildings.

Studying installed equipment capacity

The major pieces of energy consuming equipment in the building should be surveyed to see if they are properly sized. Avoid lengthy sizing calculations in an energy audit, but consider rough estimates.

Begin to redefine requirements the building must meet, such as comfort standards, amount of outside air, and outdoor conditions for which the equipment was sized. These design conditions have changed in recent years with the rising concern for energy. Likewise, equipment design conditions based on tougher use patterns than may ever occur offer a chance for down-sizing fans, pumps and other equipment, as shown even by a rough calculation. An instance is a building with an installed lighting capacity of 5 watts per sq. ft., when other similar buildings require only 2.5 watts. **A copy of the original engineer's**

Monthly Energy Costs

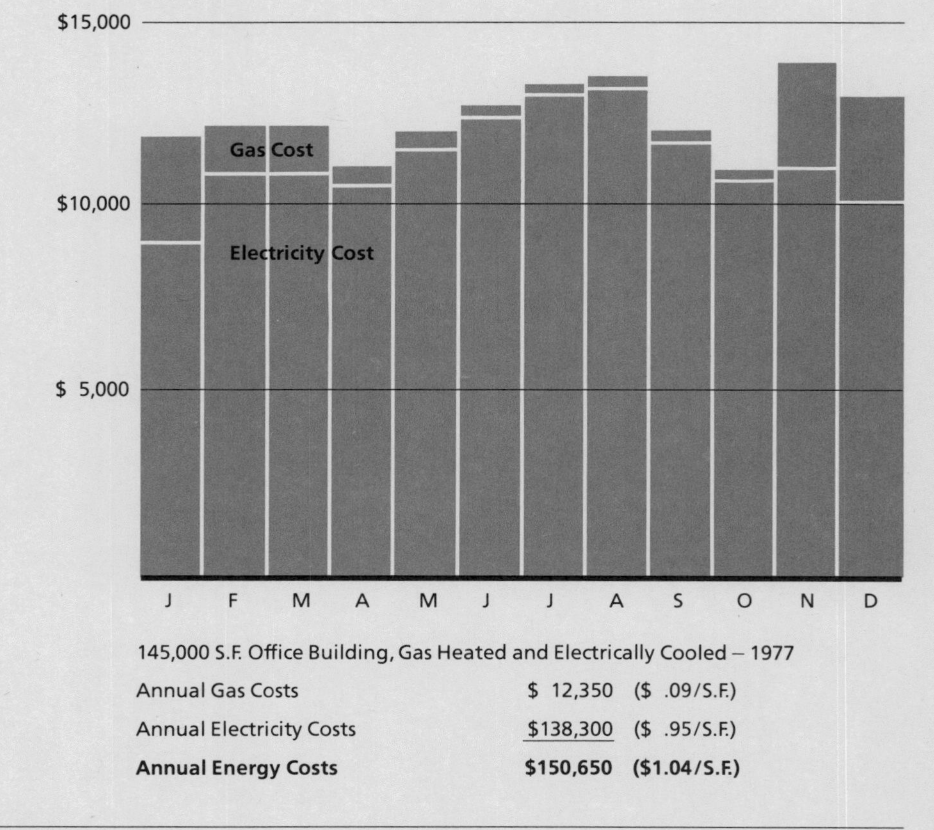

145,000 S.F. Office Building, Gas Heated and Electrically Cooled – 1977

Annual Gas Costs	$ 12,350	($ .09/S.F.)
Annual Electricity Costs	$138,300	($ .95/S.F.)
Annual Energy Costs	**$150,650**	**($1.04/S.F.)**

Figure 3-2

design calculations can be very useful in reducing the effort required, but may be hard or impossible to obtain.

Studying peak energy demand

Peak energy demand is the highest rate of energy use during the billing period, usually a month. Most utilities measure "integrated" demand. This is simply the amount of energy used during a specific time interval, usually 15 or 30 minutes, divided by the time interval in hours. For instance, if the building's equipment consumed 200 KWH of electricity during the peak 15 minute use period, its integrated demand would be 800 KW. (200 KWH/ per ¼ hr.).

A study of peak energy demand discloses excessive or unusual patterns of peak energy use and offers a chance to reduce these peaks. The point of peak reduction is to save the owner money by reducing the electric demand charge.

The purpose of the demand charge is to reduce a need for additional electric generating plants. Peak reduction is unlikely to save much energy at the building site since the need for energy is not eliminated, merely delayed and leveled out.

Once energy planners have the installed capacity of the building's equipment, the actual peak from the utility bills, and a calculation of what it should be, they can compare potential savings in peak energy demand charges. Discrepancies will warrant a chance for a closer look into possible savings.

To quickly determine the rate of energy use for individual pieces of equipment, shut off these pieces at the site and look for any change in speed of the disc in the electric meter. By applying conversion factors available from the electric utility and by timing the reduced rotation, the instantaneous rate of electric use can be calculated.

Monthly Electric Costs

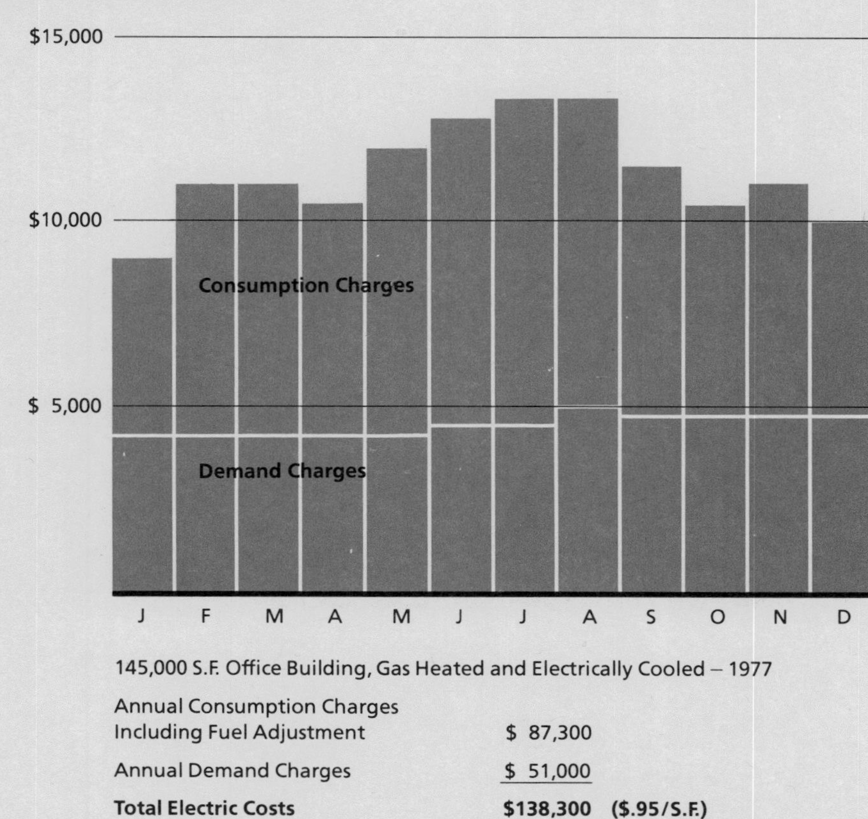

145,000 S.F. Office Building, Gas Heated and Electrically Cooled – 1977

Annual Consumption Charges Including Fuel Adjustment	$ 87,300
Annual Demand Charges	$ 51,000
Total Electric Costs	**$138,300 ($.95/S.F.)**

Figure 3-3

Actual monthly peak electric demand (not "billing demand") can be plotted for comparison (see **figure 3-4).** Months that are of interest can be explored further using **hourly energy use data** available from utilities, such as the G-9 chart.

Plot the high and low monthly temperatures on the same graph to see if there is a relationship between outside temperature and peak energy demand.

In this case, the building's electric company bases its demand charge on a **"billing demand"** which is the highest of (a) 95% of the previous summer peak demand, (b) 60% of the previous winter peak demand, or (c) the appropriate percentage of the actual demand for that month.

Conclusions emerge by reviewing the **figure 3-4 Monthly electric demand** chart:

- Actual demand fluctuates about 15%, whereas billing demand only fluctuates 5% after being set. The time at which billing demand is set is, therefore, important. There is no cost advantage to reducing the peak the rest of the year. The 1978 August new peak was expensive and will remain so until a lower summer peak is established next year.
- Peak demand was higher in 1978 than in 1977, because the 95% peak points moved up after the August peak. Perhaps the cause was higher occupancy.
- The winter peak is no problem.
- Monthly peaks are more erratic than temperatures. Perhaps this is due to several pieces of equipment coming on together.

Graphs of energy demand and temperatures are usually hard to relate. Basic relationships should stand out, however, such as rises in heating energy when temperatures go down for weather-sensitive buildings. Yet the graphs may show no correlation because of system response such as reheat, operations procedures, or control systems insensitive to the weather. Still, in most buildings, "graphing" can be a quick way to identify potential improvement areas.

Energy use can be charged to major pieces of equipment or building systems which contribute to the building's peak demand. This peak energy use can be plotted, as shown in the graph of **estimated major contributors to peak demand (figure 3-5).** If a building is designed with a large central HVAC system and runs to a predictable schedule, then **much of the building's equipment, such as lights and fans, may operate at a constant rate.** This makes it easy to estimate this energy consumption. The "draw" or energy consumption per unit of time for constant speed equipment may be read from the equipment nameplate and plotted on the demand graph as a straight line. If elevators and water heaters draw electricity at capacity at time of peaks, then their nameplate rating can also be directly added to the chart.

(Although actual draw often differs from nameplate ratings, especially for elevators, the nameplate rating provides a good approximation for use in the audit.)

Energy remaining above these lines will be due mainly to HVAC and other **loads which vary** throughout the year.

Operation of noncritical equipment should clearly be delayed during periods of peak demand to allow for **"peak shaving."** This may be done by methods ranging from using building operators to simple clock timers and complex computers.

Figure 3-5 shows that lighting is a major contributor to the peak. As such it

Monthly Electric Demand
Monthly Peak Demand; Billing Demand, High and Low Temperatures

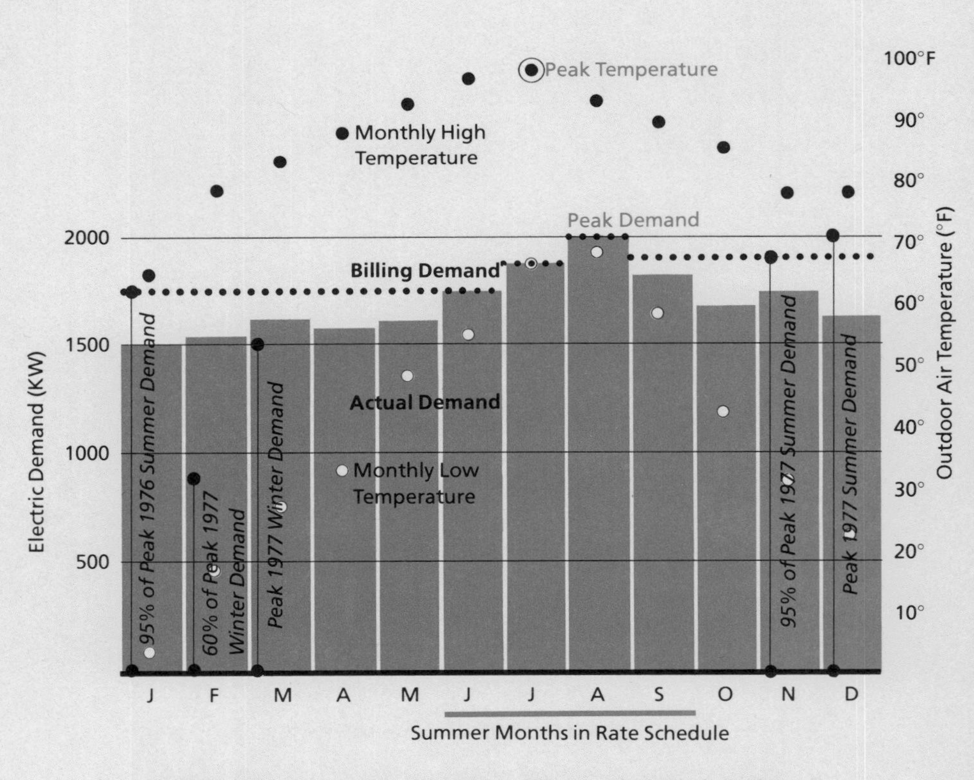

Figure 3-4

is an opportunity for reducing demand charges as well as consumption charges. Also, cooling due to the heat of lighting further adds to the peak.

The electric draw of constant flow pumps and fans is independent of outside weather conditions.

The draw characteristics of refrigeration compressors depend on the type of compressor control. If the compressor cycles on and off, the draw goes up only slightly with higher outside temperatures. The percentage of running time is the major variable. By contrast, a compressor with a built-in unloading system tends to operate constantly, but power input varies in proportion to the building cooling load.

Large energy users often have a recording demand meter installed. This provides detailed information about the building's peak energy demand as it records electric consumption at 30 minute time intervals throughout the day. Recording demand meters which produce a graphic display of the month's demand are the most useful to the energy planner. Newer magnetic tape meters known as billing tape recorders (BTR) are more accurate and useful than the older meters, but are not as good for an energy audit unless equipped with a graphic display.

For example, note the following charts produced by a General Electric G-9 recording demand meter which displays the amount of electricity used every 30 minutes over a 32-day period.

By applying the appropriate multipliers obtained from the utility company, one may use the chart to calculate the average peak demand that occurred within each 30-minute segment of time. One may then locate the segment in which the peak occurred each month to find out how the building was operated. Based on this ready-made profile of energy use and operation, the energy planner can draw a theoretical plot for a typical day, and over-lay each day's actual plot to note any discrepancies. By comparing the two curves, one may identify opportunities for improvement.

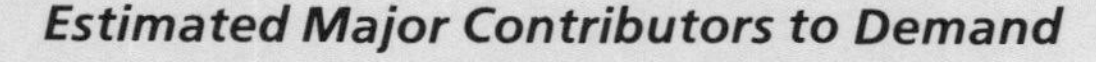

Estimated Major Contributors to Demand

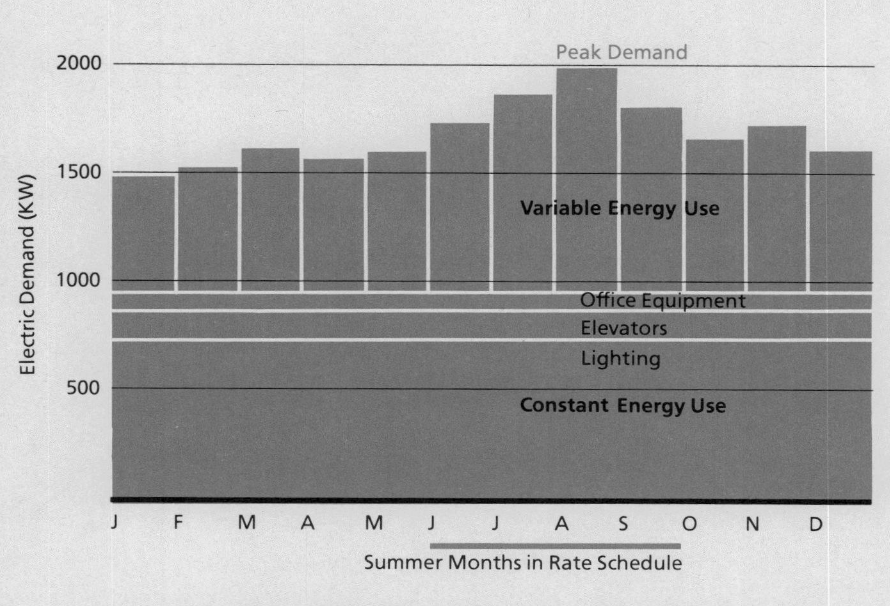

145,000 S.F. Office Building, Gas Heated and Electrically Cooled – 1977

Figure 3-5

For example, the **G-9 charts (figures 3-6, 3-7)** yield the following data about the building's energy use:

It is not actually operated on a 5-day basis, but also part-time on Saturdays and at times on Sunday. On one occasion, nearly full capacity was run all night **(figure 3-7)**.

Winter peaks **(figure 3-6)** are far lower than summer peaks **(figure 3-7)**.

Peaks occur around and immediately following lunch.

The building is started up about 5:00 to 6:00 AM and shut down around 8:00 PM. Shut down times vary. Perhaps shut down could be accomplished more promptly.

There is much low intensity use during hours when the building is supposed to be "off." The energy planner should find out why.

These charts indicate large amounts of energy that are low in intensity, but used even when the building is unoccupied and theoretically turned "off." Only a careful search late at night discloses that toilet room lights, electric transformers, corridor lights, equipment for communication systems, elevator motors, exhaust fans, business equipment and coffee makers use energy 24 hours a day. These may not be major consumers and sometimes cannot be turned off without difficulty, but all have **a high load factor.** Big savings may be achieved simply by finding a way to turn some of them off.

A **"spike" of peak demand** (where energy consumption shown on the G-9 chart suddenly extends far beyond the other normal lines) should also be accounted for and reduced. In the case of the rate schedule illustrated in this book, a single high spike sets the energy demand charge for a whole year. By and large, the less often the spike occurs, the easier it is to reduce.

Chart from Recording Demand Meter (G-9) – February

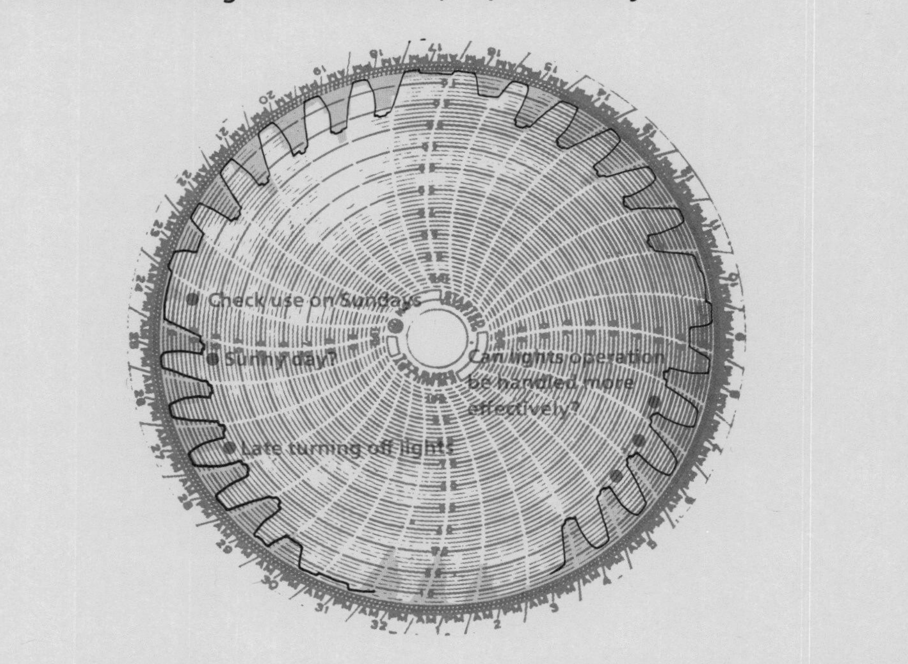

Figure 3-6

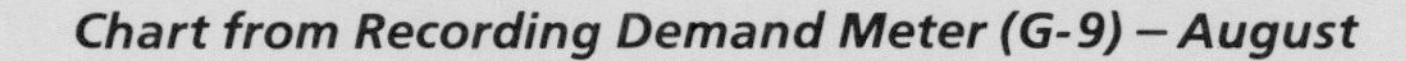

Chart from Recording Demand Meter (G-9) – August

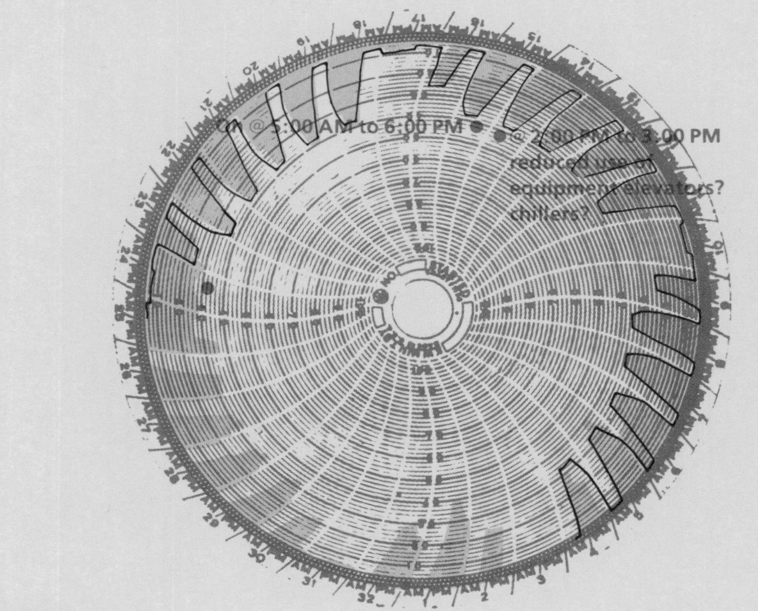

Observations:
Can lights and equipment be started up after 5:00 – 6:00 AM?
Can lights and equipment be shut off before 8:00 PM?
Why does equipment drop off at 2:00 – 3:00 PM?
Why was building left operating all weekend on days 17, 18, 19?
What consumers cause the low intensity energy use when building is theoretically "OFF"?

Figure 3-7

Studying energy consumption

Studying energy consumption from utility bills helps locate excessive or unusual patterns of energy use and provides the chance to reduce this use. The building owner's incentive to do so may be to reduce costs, use less energy to avoid shortages or meet an energy budget.

The procedure is like that used for studying peak demand. **Plot and compare the monthly energy consumption** figures for each energy source from past bills, starting with electric. Plotting more than one year will provide a better basis for comparison **(figure 3-8)**. Plotting average monthly temperatures will yield added information.

From the **monthly electric consumption** and average monthly temperature chart **(figure 3-8),** observe that:

- Because consumption is so low, either the building was only partly occupied, was operated very well, or operated fewer hours from January 1975 through at least July 1975.
- Consumption seems to rise in winter and summer and drop in fall and spring, but not substantially. Hence the building as operated was not very sensitive to weather.
- June and July 1977 were unusually high. One must ask, why?
- Consumption rises each year. Perhaps equipment, operation, procedures, central calibration or system adjustment are deteriorating.
- November 1977 consumption is unusually high, due perhaps to operation changes or malfunction.

Irregularities generally result from inefficiency in four areas; erratic operation procedures, deterioration of equipment and controls, changes in building use and unusual weather. Clearly, gross cause/effect generalizations are risky and before firm conclusions can be drawn a more detailed study is needed than is usually found in an energy audit.

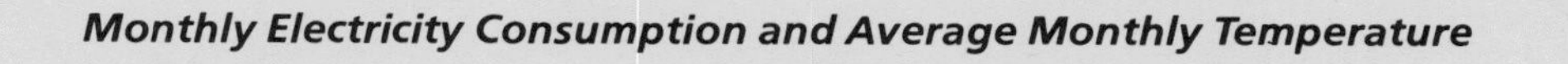

Monthly Electricity Consumption and Average Monthly Temperature

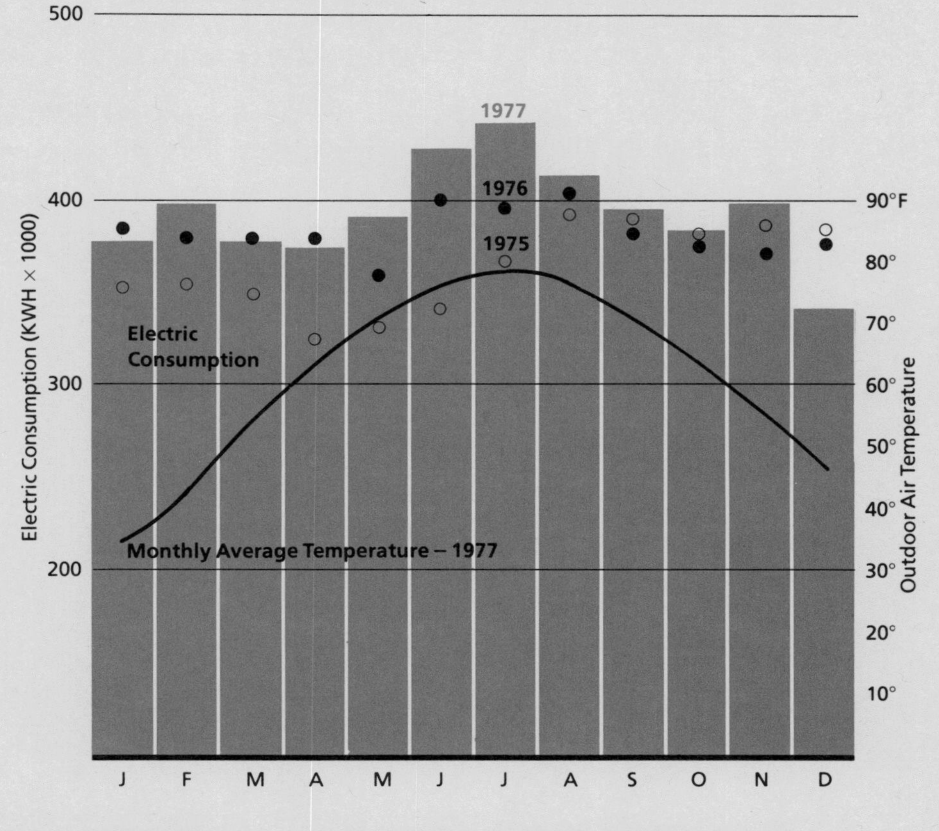

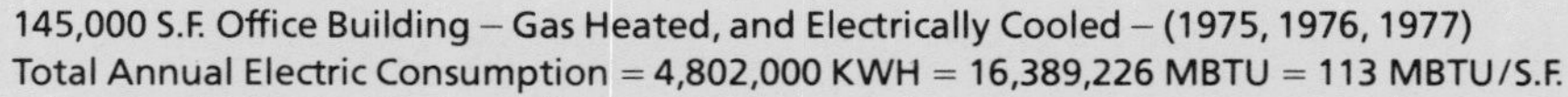
145,000 S.F. Office Building – Gas Heated, and Electrically Cooled – (1975, 1976, 1977)
Total Annual Electric Consumption = 4,802,000 KWH = 16,389,226 MBTU = 113 MBTU/S.F.

Figure 3-8

Plotting **monthly gas consumption** and average monthly temperatures **(figure 3-9.)**

This graph shows that:

Gas consumption is sensitive to outdoor temperature.

In November and January energy use rose drastically while temperatures dropped. There may have been serious malfunctions and extra operation errors in those months.

A monthly consumption estimate of non-HVAC items will be the same each month if an average of 30.4 days per month is used. Non-HVAC equipment surveyed can be plotted as horizontal straight lines as in **figure 3-10.** Energy used as shown above these lines is mainly due to HVAC energy consumption, but also includes elevators and other equipment that varies in operation. As a cross check on this HVAC estimate, use a rough equivalent full-load hour calculation, as discussed later in this chapter.

Figure 3-10 shows the relative sizes of the largest constant rate individual consumers of electricity. That is where one should focus the effort. HVAC consumption can then be broken down by identifying the constant cooling load due to lighting whenever heat produced by the building's internal load is greater than the heat required to balance outdoor ambient conditions.

From this graph, one may see that if HVAC consumption were cut in half, electrical energy would probably drop by less than 15%. If the HVAC electric load due to heat produced by lighting were, as it appears, half of the HVAC energy component, a 50% reduction in lighting would sharply reduce not only HVAC energy use,but lighting energy use as well.

Similarly, gas consumption can be broken into components. This is more difficult since gas use tends to fluctuate quickly in response to weather and therefore has few constant load characteristics.

The foregoing energy audit methods are neither complete nor conclusive. Data can be plotted in many ways to test conclusions. In the limited form of the audit there is much room for inaccurate interpretation and false conclusions. Final recommendations should be based instead on a more detailed energy analysis.

Monthly Gas Consumption

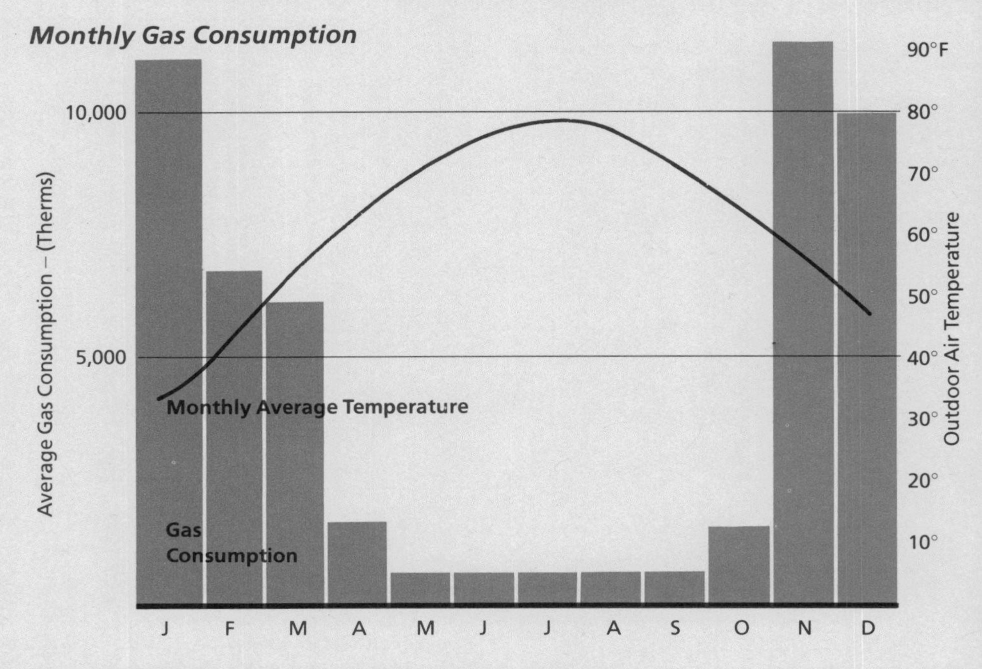

145,000 S.F. Office Building, Gas Heated, Electrically Cooled
Total Annual Gas Consumption = 14,980 therms = 1,498,000 MBTU = 10.3 MBTU/S.F.

Figure 3-9

Electricity Consumption with an Estimate of Constant Rate Non-HVAC Components

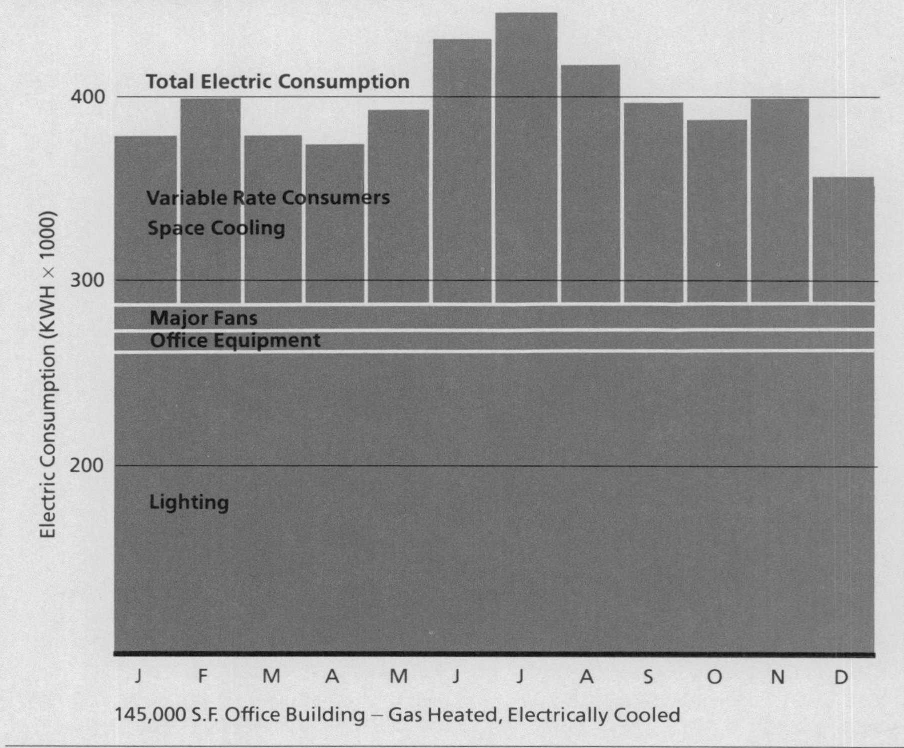

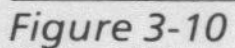

145,000 S.F. Office Building – Gas Heated, Electrically Cooled

Figure 3-10

Performing the energy analysis

The ***energy analysis is a more comprehensive study*** *of the building's energy use than is the energy audit. It is done in sufficient detail to let the energy planner definitely recommend modifications to improve the building. These recommendations call for an understanding of how, when and where energy is used, so that modifications may be reliably evaluated.*

Whereas the audit was based on existing facts of energy use, that is, utility information, the ***analysis relies on the energy planner's professional judgment*** *in interpreting and mathematically quantifying information including operation, equipment performance and systems response. By simulating the building's energy performance, the energy planner extends the limited information provided by the audit with more detailed information to the point where opportunities for improvement can be accurately identified and evaluated.*

The analysis includes:

- *Detailed study of the building, its operation and use, its environment and its equipment.*
- *Calculation of loads the building and its equipment must meet.*
- *An estimate of energy used by individual pieces of equipment in response to loads.*
- *Calculation of costs to provide this energy.*
- *Study of how the building performs in non-energy related areas.*

An energy analysis follows these steps:

- *Collecting data (see discussion in Energy Audit).*
- *Estimating energy performance.*
  - *Estimating HVAC energy use.*
  - *Estimating non-HVAC energy use.*
  - *Totaling HVAC and non-HVAC energy use.*
  - *Estimating energy costs.*

The energy analysis gives the energy planner a detailed picture of energy consumed by each piece of equipment in proportion to total energy use, as well as the effect of weather changes, operation and use patterns, plus resulting costs.

The energy planner should verify these results by comparing them to past utility bills or records for similar buildings. Conclusions should be reviewed with the owner before further work is done in identifying and evaluating opportunities. The **owner must understand results and assumptions** on which the analysis is based since they determine steps to take to improve the building's performance.

Developing the analysis procedure

Results of the energy audit and review with the owner serve to pinpoint areas of investigation for the energy analysis. Since scope of the analysis can vary widely, two tasks must precede it:

- *Determine the degree of effort and accuracy required from the energy analysis.*
- *Choose the type of estimates to be used.*

Review the energy audit's predictions of possible savings and check to see if they can be done within the project's budget and schedule.

Estimate accuracy and depth required to evaluate opportunities. In any project, several opportunities are clearly worth pursuing; most will already have surfaced in the energy audit. To evaluate others will call for more detailed analysis since they may be highly interrelated with other factors influencing the building's performance.

As an example, the impact of a more efficient furnace may be easy to evaluate accurately. On the other hand, adding solar shading devices will modify the building's HVAC equipment's required capacity and performance, the building's appearance, internal lighting levels and comfort conditions. What is more, all of these change hour-by-hour with the sun's movement. Only a detailed analysis will provide the facts. The most critical task may well be assessing the effect of shading devices on the building's appearance, not the estimate of furnace efficiency.

The degree of detail of analysis should reflect the level of savings possible for the item being analyzed. If elevator operation seems inefficient but accounts for only a small part of the building's total energy use, there is no point in making a detailed analysis of it. Likewise, why concentrate on an item if little can be done to change it?

On the other hand, the energy planner may want to study major pieces of equipment such as chillers thoroughly and even monitor them hour-by-hour. Chiller performance may be difficult to estimate since it responds to many variables, and manufacturer's data as to their response to other than peak conditions may be inaccurate or lacking. The effort may be worthwhile.

Even though it should not cover every piece of the building's equipment at the same level of detail, **the energy analysis should not overlook any equipment.** Assumptions can be used to fill in data gaps regarding equipment that need not be studied in detail, so long as impact of the small item on the larger items is known.

A more complete study will allow for analyzing opportunities that may be uncovered only after the energy analysis is complete. If information is missing, the energy analysis may need to be completely redone to include it. Each step and assumption in the analysis should be thoroughly documented so calculations can be revised as added opportunities emerge for improvements.

For example, if replacing windows to reduce infiltration looks viable at the end of the analysis, but the effect of infiltration was not first included, there is no way to calculate the possible savings.

Reviewing the project with these concerns in mind will allow the energy planner to decide how accurate and how detailed to make the energy analysis.

Collecting data for the energy analysis

Much of the data required for the energy analysis will already have been gathered for the energy audit. Added data will need to be more detailed, more reliable, and include a broader range of concerns, such as a detailed look into comfort conditions in the building.

During the data collection effort, chances for energy related and non-energy related improvements will become evident. Non-energy related opportunities are **a bonus to the owner** and may suggest design improvements which can be incorporated in the final solution. Users may have no interest in the owner's goal of saving energy, but they will be very much concerned with the non-energy related conditions which result; **do not overlook these broader objectives** through preoccupation only with energy performance. An improved environment should be expected to result.

Information used in the analysis must be accurate, and any assumptions regarding use and operation must be documented. For instance, if HVAC equipment is assumed to operate 18 hours a day, seven days a week, but really operates 12 hours a day, five days a week, then calculations will be very inaccurate. If indoor temperature for cooling is assumed as 78°F, but really operates at 72°F, results again will be misleading. If there are operable internal shading devices, are they to be taken as open, closed, or 50% closed?

If the building's energy use and costs during a specific year's operation are to be compared with actual records, it is important to define the years, since weather, operations, use characteristics and energy costs differ with various base years.

Once selected, all data must be related to that **base year.**

The energy analysis will link the energy consuming items. Each project must make certain assumptions whenever data is not available. Data and assumptions should be reviewed with the owner before the energy analysis is begun because the owner may have fresh information or disagree with some of the assumptions. Also, changes may be pending, such as newly signed tenants with large computer operations that require 24-hour a day use. All assumptions should be accepted by the owner in writing before the analysis gets under way.

Organizing the data simplifies analysis. Keep data together with recorded sources and assumptions in a notebook which can be updated and used as a common reference. If the energy performance estimate must be done on a zone-by-zone basis, as is usually the case, it helps to have the information broken down and organized by zones. This tends to apply to information on materials and various wall, roof and floor sections, inventories of people, lights and equipment (see **figure SP-6**).

Additional data over and beyond what was needed for the audit is available from:

The building

Any sources beyond drawings may be difficult to get. Yet in the analysis the added information as to how the building was actually constructed may be critical, such as:

Drawings (as-built, if possible)
Specifications
Shop drawings
Addenda

Building's architect and design engineer

Talk to the building designers. **Assumptions** used during the building's design may have changed or proved inaccurate and modifications may now be made. The engineer's design calculations, if available, can be useful for determining equipment design capacity.

References

References can lead to information on material, equipment and methods. See the reference list in the back of this book.

The building's original contractor

Consult with the builder to see if plans of record accurately show the way the building was built. Check to see if there were any unresolved problems.

On-site observation and testing

Other data can be obtained by first-hand observation. For example, how often are entrance doors opened and how long do they stay open as people enter?

On-site observation will reveal the condition of the building. Look for leaks, signs of moisture penetration or items in need of repair. Check to see how well dampers, doors and windows fit their openings and how tightly they close. Review caulking and weather-stripping. Sometimes controls become disconnected, strainers and filters become clogged.

Environment

To information from the audit, add weather data, including solar radiation.

TRY (Test and Reference Year)

Future standard weather data will be the ASHRAE TRY (Test and Reference Year) data, now available for many locations in the form of hour-by-hour magnetic tapes. Currently, this data must be processed by computer into a format suitable for energy estimates using the bin method, although this data is expected to be printed soon.

ASHRAE Handbook of Fundamentals and Systems Volumes

Hourly dry-bulb temperature occurrences are listed for major cities in the 1976 ASHRAE Systems Volume (ref.).

Percentage sunshine data are listed in **ASHRAE Transaction** 1974, Volume 80, Part II (ref.). Maps with similar data have been published by the U.S. Department of Commerce, Environmental Data Service. (Some disagree as to definition and accuracy of the above data if used to predict solar heat gain.) On-site solar radiation measurements should certainly be used if available. If the energy estimate is being done to compare alternatives, however, errors in atmospheric correction factors will be minor so long as the same assumptions are used for all comparative estimates.

The standard industry source of solar heat gain data is the ASHRAE Handbook of Fundamentals (ref.).

Air Force Manual 88-8

Hourly dry bulb and wet bulb temperatures organized into three eight-hour segments of the day for each five-degree temperature range (bin) are listed in the Air Force Manual 88-8, Chapter 6, "Engineering Weather Data." (ref.).

This type of data is essential for all bin method energy estimates.

U.S. Coast and Geodetic Service survey maps

The altitude and latitude of a site can always be determined where this information is not otherwise available. Local suppliers of these maps are listed in telephone yellow pages directories.

On-site observation

Site data such as location and size of surrounding trees, buildings, hills or reflective surfaces, is best obtained from on-site observation as recorded on site-drawings. Temperatures can be monitored on-site for comparison with weather stations.

Equipment (see figure SP-9)

An inventory should be made of any equipment that consumes energy. The level of detail of the inventory is determined by the purpose of the analysis and time available. For convenience, building equipment can be broken down into categories, as illustrated in **figure 3-1**.

Each equipment category is made up of several types and combinations of equipment. For example, a cooling system may include compressors, pumps and fans. They will be electric-powered or fuel-burning. Some will consume energy at a constant rate, others will modulate their rate in response to loads. Following is a summary of information needed.

Electrical equipment

Location of and area served:
Nominal size: usually given in terms of horsepower or watts. Operating conditions: design and actual. This describes the job the equipment was intended to do and what it is actually doing. Included here would be such information as fan static pressure and cfm; pump gpm and pressure drop; cooling and heating equipment entering and leaving air or water temperatures, etc.
Voltage: Actual and nominal.
Amperage: Full-load and actual.
Amount of heat given off: amount of heat and its point of impact on HVAC equipment. Rejected waste heat and heat contributed to HVAC load or from useful output should both be noted.

*Lighting fixtures: (see **figure SP-10**).*

Type, location and number:
Lamp wattage, ballast wattage, condition and dimmer characteristics.
Amount of heat given off: amount of heat and its point of impact on HVAC equipment.
Lighting task: light required, sources of glare or veiling reflections.

Fuel-fired engines or boilers

Nominal size: output, input, or model number.
Operating conditions: design and actual. Includes: fuel input rating, design load; equipment output rating; and efficiency.
Amount of heat given off: amount of heat and its point of impact on HVAC equipment.

Some manufacturers will have part-load performance information; most do not. More of them will have it due to current emphasis on equipment operation costs. Ask them for certified part-load test results, or test on the job.

Building's operators and maintenance crew

Buildings are often maintained by outside companies and may thus not have an in-house operator. The persons involved are the **most immediate and direct source of information and ideas** to improve the building, but they usually lack the technical breadth of the energy planner. If the latter can place what he learns from the building staff into a proper technical context, the analysis will move forward more smoothly.

Operating methods (see figure SP-11)

Include past, present and future.

Building's original design program
Design engineer's assumptions and calculations
Operation manual should describe hours and methods of operation
Time clock settings will show when equipment starts and stops
Controls contractor will know settings
Owner and management staff
Building users know how they use the building and its conditions
"Sign-in" sheets indicate building occupancy
On-site observation will verify information

To see how a building and its equipment are actually used and operated, all the above sources are useful, but on-site observations and conversations are best. **Savings can result from matching building and equipment operation schedules to the building's actual use schedules.** Most energy analyses are done so quickly that they do not allow building observation during all seasons of the year. In the audit the energy planner may have taken the operator's or owner's description of the operating hours as accurate; in the more detailed analysis, time should be taken to observe a typical daily start-up and shutdown, as well as night-time and weekend operation.

The energy planner should determine when spaces in the building are actually used, by how many people and for what activities. Average and peak occupancy and future occupancy changes are very important in uncovering improvements. If the building is to be modified for a change in occupancy, the new design can be planned for energy-efficiency at no additional cost. Maintenance and cleaning procedures should be looked into after the building occupants have left. Check to see what operates in the building during the "off-hours."

It is good practice to field check control settings personally since the control system may not be properly calibrated. Determine actual operation schedules accurately since they have a major impact on energy use and the energy use estimate. Be forewarned that there are usually several different versions of how a building is operated.

Energy sources

Local utility companies

Basic rate schedule information from utilities and fuel suppliers should have been obtained and examined as part of the audit. If the analysis is to be projected for future years, estimate future rates and energy availability.

The energy analysis should provide figures in a format that can be applied to the **appropriate rate schedule.** For instance, if the electric company bills each month, monthly consumption and demand figures will be required from the analysis to calculate energy costs.

Suppliers will have several rate schedules, based on types of users and end use. The building may fall into a commercial user classification, but have different rate schedules applied to its outdoor lighting, its domestic hot water and its general consumption. Each item will need to be kept separate in the energy estimate. In some cases, a negotiated rate may apply rather than one of the supplier's standard rate schedules, especially in case of a large complex of buildings or an industry. One may then need to know how reductions in energy consumption or demand will affect the negotiated costs.

Applicable billing periods

The billing period may be a year, a season or a month. In addition, companies will increasingly charge **time-of-day rates,** with higher prices during the supplier's peak load period. The energy estimate may have to separate totals for each time-of-day billing period.

Minimum rates, "ratchet" rates, additional charges and penalties

Ratchet rates are often used to bill electric demand. If maximum annual or seasonal demand controls the bill, rather than actual demand during the current billing period, this maximum demand is referred to as a ratchet. Only the annual or seasonal demand may then need to be estimated. Additional charges and penalties are discussed under "estimate energy costs" later in this chapter.

Appropriate billing units

Each charge may be based upon a separate unit such as gallon, ton, cubic foot, kilowatt or kilowatt hour.

Utility rates

Since the components of utility rates are mostly regulated, projections of future rates are whatever the building owner and energy planner wish to assume. For credibility, the energy planner should check, in order of priority, with the following:

Building owner
Local utility and supplier
State energy commission

At the same time, extrapolate the owner's historical energy costs.

Comfort and other factors

The energy planner must view the energy analysis in context of other aspects of the building's performance such as comfort, appearance, function, environmental impact and maintenance. Personal observation by the energy planner and by those who use the building regularly are the best source.

Uncomfortable rooms and other problems noted in the walk-through should now be studied in greater detail.

The energy planner may wish to measure some basic conditions such as air and wall surface temperatures, humidity, air flow rates or lighting levels if there is doubt as to the cause of reported comfort problems. Small scale plans of the building and photographs may be used to record observations.

Surveys of the building's users are the best way to a more complete picture of the building. If the questionnaire is subjective, users should be assured the data will be kept confidential.

Comments from users often show that they feel strongly about their environment. To avoid conflict, the questionnaire should point out that its aim is to help improve the building's comfort level as well as to increase energy efficiency.

In general, questionnaires should be phrased so as to require yes-or-no answers, or very short responses. If the survey form can be completed in ten minutes or less, it has a better chance of being filled out.

Numerous reports of uncomfortable conditions in any part of the building indicate a serious problem. The promise of improved job performance is strong motivation for employers to improve comfort.

Estimating energy performance

This chapter divides the estimating procedure into three identifiable parts.

*A Compare **equipment capacity** with requirement*
*B Calculate **peak energy demand***
*C Calculate **average energy consumption***

This procedure is applied to HVAC energy use and, separately, to non-HVAC energy use. The two totals are then added to obtain the building's total energy use. Various sub-totals may then be developed and shown on graphs to convey specific information more conveniently.

Part A is often overlooked. It **compares equipment capacity and actual job requirements.** This comparison helps identify opportunities for saving energy by eliminating wasteful excess capacity. For example, providing an average 120 foot-candles of lighting when 60 would suffice, or constantly circulating 20,000 cfm of air when 15,000 cfm would handle the peak cooling load, is wasteful. Yet this may not appear as such without deliberately comparing capacity and actual requirement.

Equipment capacity must also be identified since energy usage of many items varies with the percent operating load, which cannot be obtained without knowing maximum capacity.

Part B is calculation of peak energy demand. It is needed only if the utility energy cost involves a demand charge. Most electric and some gas utilities include a demand charge in their rate structure. If the demand charge is on a monthly basis, peak energy demand must be estimated separately for each month.

Note that demand rate is set by the simultaneous consumption of many items. For example, peak demand of the HVAC system may not coincide with peak demand of elevators. Peak demand must be calculated for the hour when overall building demand reaches its peak.

Part C is calculation of average energy consumption. This too must be coordinated with the utility energy rate structure. If billing is monthly and cost per energy unit varies with either consumption or demand, energy use must be summarized separately for each month, as shown in the sample problem.

A building is usually served by a single electric meter which measures the combined power usage of all electrical equipment. For calculation and review, it is good to separate energy calculations into the major components of energy use: HVAC, lighting, vertical transportation, domestic hot water and process loads. HVAC calculations deserve a lot of care since the HVAC operation is usually sensitive to more dependent variables than any other part of the building. That is why most of this discussion centers on HVAC.

Results of the separate estimates are easily added to obtain the total building estimates, as shown in **figure 3-11, Parts of the energy estimate.** The same process can be used to obtain energy usage and peak demand for any energy source.

Estimating HVAC energy use

Most of the bin method calculation as described in this book is spent on developing "profiles" of peak demand usage and average energy consumption. This is done by calculating system energy consumption under each of several outside weather and building operation conditions, then plotting results at each condition or "point" to obtain the curve or "profile."

Profile point calculation is similar to peak design calculation. It differs in that outside weather conditions are not the annual extremes; and for the "average" profiles you must estimate average solar and internal loads, instead of the maximum values used for design purposes.

Three or four points are usually required for the average load profile: another three or four are needed when a peak demand load profile is required. The calculations for each point are similar, with many items remaining the same for all points.

These repetitive types of calculation can be simplified by use of a programmable calculator.

The **procedure for calculating each profile point is illustrated in its simplest form in Figure 3-12** and consists of:

Parts of the Energy Estimate

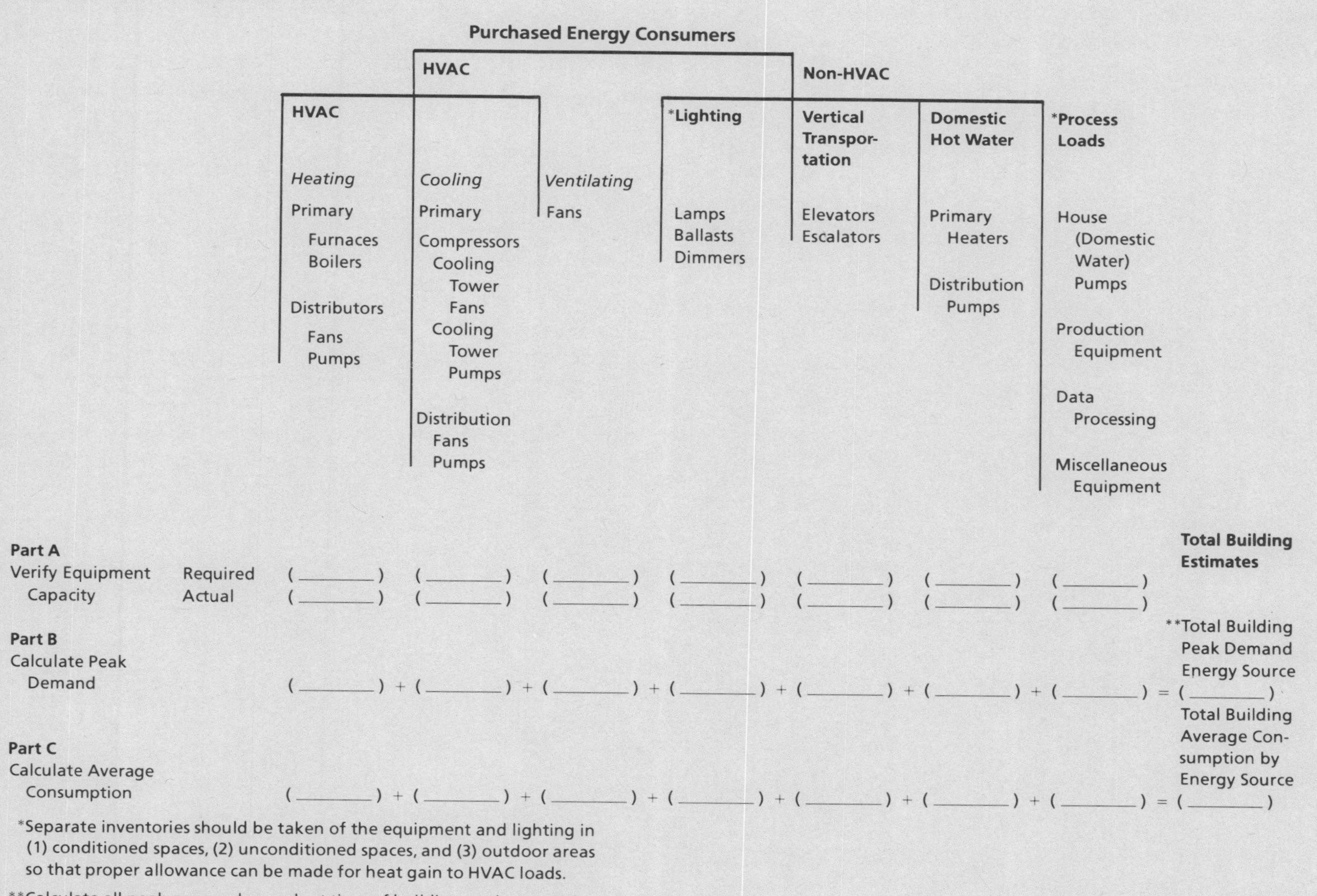

*Separate inventories should be taken of the equipment and lighting in (1) conditioned spaces, (2) unconditioned spaces, and (3) outdoor areas so that proper allowance can be made for heat gain to HVAC loads.

**Calculate all peak energy demands at time of building peak, not at the time each item individually peaks.

Figure 3-11

Establishing the type of profile point, that is, peak demand or average energy consumption.

Establishing the internal load due to operation consistent with the type of profile point. For example, a peak cooling load will occur during full occupancy whereas a profile point for average consumption may combine internal loads during both part occupancy and full occupancy.

Establishing outside weather conditions at this point. Inside room conditions are the same for all points, unless you are reflecting a change in room temperature set-point from summer to winter.

Calculating thermal loads, system response and hourly energy consumption of each item of equipment. This energy input can be summed up and plotted as a total system input, or else plotted separately to track an individual item.

Looking at it another way, an HVAC system may be seen as having a separate meter on each piece of energy consuming equipment. The profile point calculations are simply an estimate of these meter readings at an operating condition similar to the profile point. To do this, one must first estimate the building thermal load at a specific set of conditions (the profile point), determine response of that particular type of HVAC system to that percent of design load, then determine energy input to the equipment as it handles its load. This operation can be simple or difficult, depending on the project.

A simple case is a building with a single HVAC temperature control zone, such as a one-room commercial building with a window-type air conditioning unit, a small multi-room office building or residence with a single-zone central system, or even a large gymnasium or manufacturing plant with a single-zone HVAC system of 200 tons capacity. Size has no effect, nor whether the system is a packaged unitary type or a custom-designed field-fabricated central type. The controlling factor is that there is only a single temperature control zone. The process is still that shown in **figure 3-12 – Calculate profile points – HVAC; single zone system.**

When additional zones are added to the single zone the possibility is introduced that one zone may require heating while another requires cooling. Combining the zones into one estimate would permit the cooling load of one zone to cancel the heating load in the other. The result would be underestimating of both cooling and heating loads. Hence zones must be computed separately, given any chance that both heating and cooling requirements may occur at the same profile point. If the project uses multiple single-zone HVAC systems (unitary or field-assembled), make a separate estimate for each zone, then add the energy consumption as shown in **figure 3-13 –**

Calculate HVAC Profile Points; Single Zone System

Calculate energy consumption at each profile point keeping track of each energy source separately

Start

Establish Type of Profile Point
(Peak demand or average consumption)

Establish Internal Load Conditions

Establish Weather Conditions

Calculate Zone Thermal Load
(Transmission, solar, internal lights, internal equipment, others...)

Calculate Zone System Response
(Fans, reheat, others...)

Calculate Zone Equipment Energy Consumption

	Elec.	Fuel
(Fans	(______)	(______)
Refrigeration	(______)	(______)
Heating	(______)	(______)
Other...)	(______)	(______)

Total Zone HVAC Energy Consumption	**Elec.**	**Fuel**
	(______)	(______)

Profile Points for Total Building HVAC

Figure 3-12

Calculate HVAC profile points – multiple zone system.

The multiple zone central system is even more complicated since the central refrigeration system (and possibly the central heating, air handling and pumping systems) are simultaneously handling the combination of many individual loads imposed by the local room temperature control zones, each of which acts independently. If there is a local fan or electric resistance type heater in each zone, energy consumption of this zone equipment should be related to the cooling and heating needs of that individual zone. Also, it could be that some thermal loads, such as ventilation air loads or return air heat gain or loss, are handled directly by the central equipment, not by local zone equipment.

Finally, the great variety of multiple-zone system control arrangements makes it harder to identify exactly how the system is being controlled and to define the corresponding system response to use in translating thermal loads into equipment loads.

For example, most manufacturers of package rooftop multi-zone units use different control schemes. Some use hot gas reheat; others compressor unloaders or multiple compressors. Outside air economizer control schemes vary widely. Even large variable volume systems are not always as efficient as claimed: some are of the reheat or dual-duct mixing type, which cuts down on the efficiency of their system response.

The correct procedure for analyzing a system with multiple zones and central refrigeration or heating equipment is shown in **figure 3-14 – Calculate HVAC profile points; multiple zone central system.** Thermal load in each zone is calculated separately; the local zone system response determines the resulting loads placed on both local and central equipment. Energy consumption of local zone equipment is then determined. After all zones are calculated, their contributions to central loads are totaled and added to the central ventilation loads. The central system's response converts these loads into requirements for each item of central equipment, and the energy consumption for each item is determined.

Once the profile points have been calculated they are plotted on the profiles, and the profiles used to determine the monthly totals.

The modified bin method for estimating HVAC energy use from calculating profile points to summarizing the monthly totals can be illustrated by the accompanying **Master Chart (figure 3-15).**

The energy consumption of non-HVAC equipment is not directly related to outdoor air temperatures and is calculated using simple uniform response type energy summaries. Because the HVAC system performance varies with outdoor air temperature, it requires a bin method type energy summary.

Calculate HVAC Profile Points; Multiple Zone System

Calculate energy consumption at each profile point keeping track of each energy source separately

Start

Establish Type of Profile Point
(Peak demand or average consumption)

Establish Internal Load Conditions

Establish Weather Conditions ——→ **Repeat Procedure for Each Zone**

Zone 1

Calculate Zone Thermal Load
(Transmission, solar, internal lights, internal equipment others...)

Calculate Zone System Response
(Fans, reheat, others...)

Calculate Zone Equipment Energy Consumption

	Elec.	Fuel
(Fans	(______)	(______)
Refrigeration	(______)	(______)
Heating	(______)	(______)
Others...)	(______)	(______)
Total Zone HVAC Energy Consumption	(______)	(______)

Plus +

Zones 2, 3, 4....

Calculate Zone Thermal Load
(Transmission, solar, internal lights, internal equipment others...)

Calculate Zone System Response
(Fans, reheat, others...)

Calculate Zone Equipment Energy Consumption

	Elec.	Fuel
(Fans	(______)	(______)
Refrigeration	(______)	(______)
Heating	(______)	(______)
Others...)	(______)	(______)
Total Zone HVAC Energy Consumption	(______)	(______)

Equals (=)

Elec.	Fuel
(______)	(______)

Profile Points for Total Building HVAC

Figure 3-13

Calculate HVAC Profile Points; Multiple Zone Central System

Calculate energy consumption at each profile point keeping track of each energy source separately

Start

Establish Type of Profile Point
(Peak demand or average consumption)

Establish Internal Load Conditions

Establish Weather Conditions → **Repeat Procedure for Each Zone** → **and then for Central System**

Zone 1

Calculate Zone Thermal Load
(Transmission, solar, internal lights, internal equipment others…)

Calculate Zone System Response
(Fans, reheat, others…)

Carry Zone Load on Central System to Central System

Calculate Zone Equipment Energy Consumption

	Elec.	Fuel
(Fans	(______)	(______)
Refrigeration	(______)	(______)
Heating	(______)	(______)
Others…)	(______)	(______)
Total Zone HVAC Energy Consumption	(______)	(______)

Plus
+

Zones 2, 3, 4....

Calculate Zone Thermal Load
(Transmission, solar, internal lights, internal equipment others…)

Calculate Zone System Response
(Fans, reheat, others…)

Carry Zone Load on Central System to Central System

Calculate Zone Equipment Energy Consumption

	Elec.	Fuel
(Fans	(______)	(______)
Refrigeration	(______)	(______)
Heating	(______)	(______)
Others…)	(______)	(______)
Total Zone HVAC Energy Consumption	(______)	(______)

Plus
+

Central System

Calculate Central Thermal Load
(Central system portion of zone loads, transmission, others…)

Calculate Central System Response
(Fans, pumps, others…)

Calculate Central Equipment Energy Consumption

	Elec.	Fuel
(Fans	(______)	(______)
Refrigeration	(______)	(______)
Heating	(______)	(______)
Pumps	(______)	
Others…)	(______)	(______)
Total Central HVAC Consumption	(______)	(______)

Figure 3-14

Equals (=)	Elec. ( )	Fuel ( )
	Profile Points for Total Building HVAC	

HVAC and non-HVAC are accordingly shown as computed separately in the master chart.

This master chart provides for both local zone and central type HVAC equipment and indicates that a separate estimate must be made for each energy source, typically electricity and fossil fuel.

Please refer to this chart during the detailed discussion which follows.

The calculation procedure is long but contains many similar steps in the various stages. The procedure in outline is this:

Estimate HVAC energy use

Part A – Compare HVAC equipment capacity with requirements.
- *Step A1 – Calculate zone cooling requirements*
- *Step A2 – Calculate central system cooling requirements*
- *Step A3 – Calculate zone heating requirements*
- *Step A4 – Calculate central system heating requirements*

Part B – Calculate HVAC peak energy demand.
- *Step B1 – Calculate and plot profile points (cooling demand)*
- *Step B2 – Calculate and plot profile points (heating demand).*
- *Step B3 – Compute monthly peak demand summary*

Part C – Calculate HVAC average energy consumption.
- *Step C1 – Calculate and plot profile points (average energy consumption during occupied period)*
- *Step C2 – Calculate and plot profile points (average energy consumption during unoccupied period)*
- *Step C3 – Compute monthly energy consumption summary.*

Part A – Compare HVAC equipment capacity with requirements

Determine actual installed capacity of the HVAC system and the minimum capacity needed for suitable comfort conditions within the building. In this context, "capacity" refers to cooling and heating equipment as well as air and water flows.

If the HVAC system is zoned, capacity data is needed for each zone, along with any installed central equipment.

Comparison of actual installed capacity with the minimum required will show if excess capacity is available. It usually is. A 10% or 15% surplus is typical and usually results in no big loss of efficiency. Substantial excess capacity should be looked into, however, because **an oversized system is inefficient** and should be modified.

Data on actual installed capacity is also necessary for calculating system response and equipment energy consumption wherever the items depend on percentage of installed capacity, as represented by actual part load operation condition.

Actual installed capacity data can be obtained from the design documents, submittal data or manufacturers' ratings The minimum required capacity is obtained by calculation, sometimes confirmed by observing how the system performs under extreme conditions.

These calculations point to the greatest loads likely to come up with reasonable frequency and duration, along with equipment capacity needed to handle them.

The choice of design cooling and heating conditions affects design loads and equipment selection. If room design conditions in the cooling season are warmer, room cooling load will be lower and the capacity and efficiency of the cooling equipment will be greater. If room design conditions in the heating season are cooler, room loads will be lower. Then, if certain types of equipment, such as air-to-air heat pumps, are involved, the heating equipment capacity and efficiency will be greater.

Changing outdoor design weather conditions likewise affect loads and hence equipment capacity and performance. If indoor and outdoor design conditions are not mandated by energy codes, it is good practice to discuss the choice of design conditions with the owner. Confirm the final decision in writing to avoid having to rework an entire estimate due to faulty communication.

Master Flow Chart for Estimating HVAC Energy Use Using the Modified Bin Method

A diagram of the master chart is included with the detailed discussion of each step to show which parts of the modified bin method procedure are used.

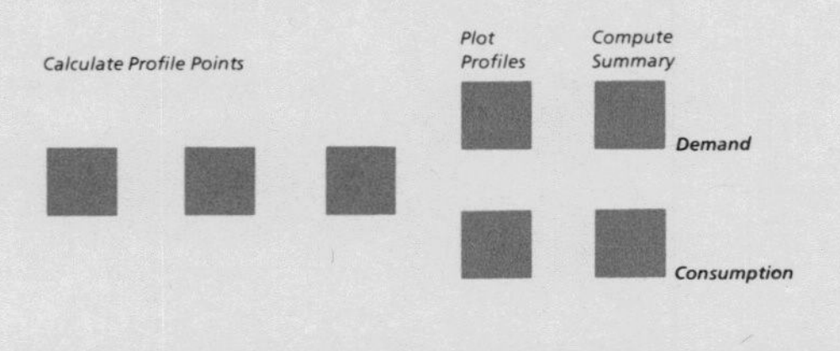

Calculate HVAC Profile Points; Multiple Zone Central System

Calculate energy consumption at each profile point keeping track of each energy source separately

Start

Establish Type of Profile Point
(Peak demand or average consumption)

Establish Internal Load Conditions

Establish Weather Conditions → **Repeat Procedure for Each Zone** → **and then for Central System**

Zone 1

Calculate Zone Thermal Load
(Transmission, solar, internal lights, internal equipment others...)

Calculate Zone System Response
(Fans, reheat, others...)

Carry Zone Load on Central System to Central System

Calculate Zone Equipment Energy Consumption

	Elec.	Fuel
(Fans	(______)	(______)
Refrigeration	(______)	(______)
Heating	(______)	(______)
Others...)	(______)	(______)
Total Zone HVAC Energy Consumption	(______)	(______)

Plus (+)

Zones 2, 3, 4....

Calculate Zone Thermal Load
(Transmission, solar, internal lights, internal equipment others...)

Calculate Zone System Response
(Fans, reheat, others...)

Carry Zone Load on Central System to Central System

Calculate Zone Equipment Energy Consumption

	Elec.	Fuel
(Fans	(______)	(______)
Refrigeration	(______)	(______)
Heating	(______)	(______)
Others...)	(______)	(______)
Total Zone HVAC Energy Consumption	(______)	(______)

Plus (+)

Central System

Calculate Central Thermal Load
(Central system portion of zone loads, transmission, others...)

Calculate Central System Response
(Fans, pumps, others...)

Calculate Central Equipment Energy Consumption

	Elec.	Fuel
(Fans	(______)	(______)
Refrigeration	(______)	(______)
Heating	(______)	(______)
Pumps	(______)	
Others...)	(______)	(______)
Total Central HVAC Consumption	(______)	(______)

Equals (=)

Elec.	Fuel
(______)	(______)

Profile Points for Total Building HVAC

or *or*

Figure 3-15

Plot HVAC Demand Profiles

Plot profile points at selected outdoor air temperatures. Separate heating and cooling profiles may be necessary if a monthly demand estimate is required.

Compute HVAC Demand Summaries

Choose higher of cooling or heating demand for each month.

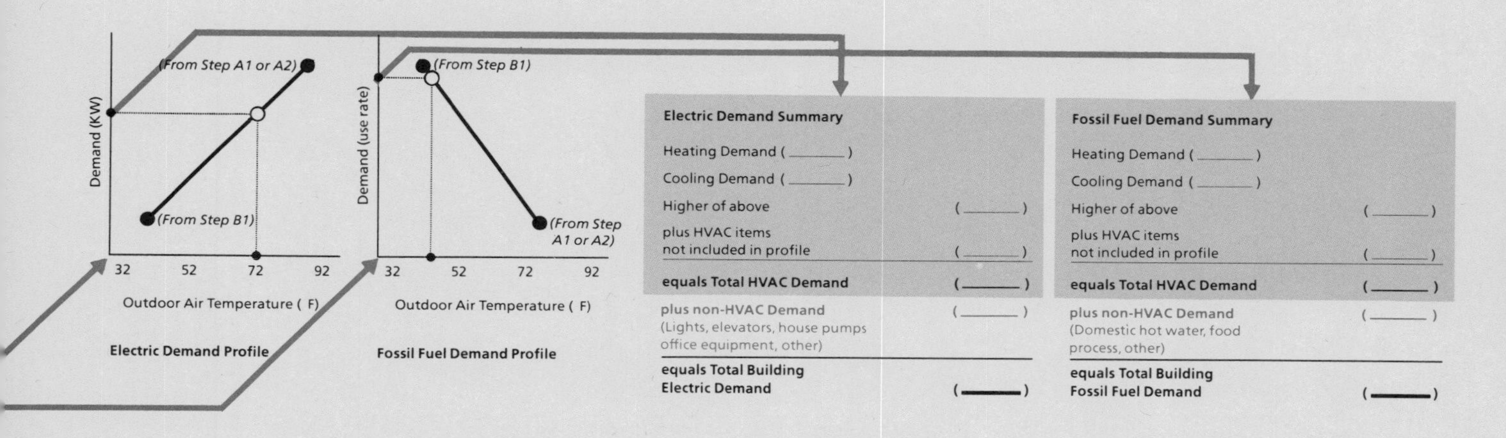

Electric Demand Summary

Heating Demand (______)
Cooling Demand (______)
Higher of above (______)
plus HVAC items not included in profile (______)
equals Total HVAC Demand (______)
plus non-HVAC Demand (Lights, elevators, house pumps office equipment, other) (______)
equals Total Building Electric Demand (______)

Fossil Fuel Demand Summary

Heating Demand (______)
Cooling Demand (______)
Higher of above (______)
plus HVAC items not included in profile (______)
equals Total HVAC Demand (______)
plus non-HVAC Demand (Domestic hot water, food process, other) (______)
equals Total Building Fossil Fuel Demand (______)

Plot HVAC Consumption Profiles

Plot profile points at selected outdoor air temperatures. Separate profiles may be necessary for each occupancy period.

Compute HVAC Consumption Summaries

Total the product of hourly consumption from the profile and the hours of occurrence from the bin weather data for all outdoor temperature bins.

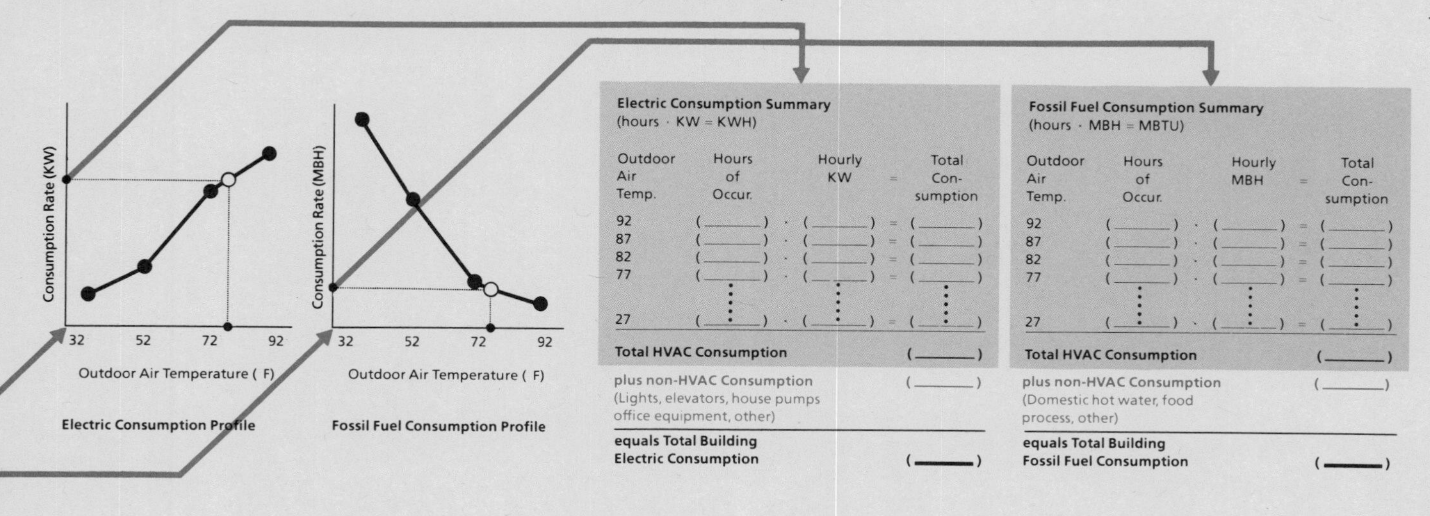

Electric Consumption Summary
(hours · KW = KWH)

Outdoor Air Temp.	Hours of Occur.		Hourly KW	=	Total Con-sumption
92	(____)	·	(____)	=	(____)
87	(____)	·	(____)	=	(____)
82	(____)	·	(____)	=	(____)
77	(____)	·	(____)	=	(____)
⋮	⋮		⋮		⋮
27	(____)	·	(____)	=	(____)

Total HVAC Consumption (______)
plus non-HVAC Consumption (Lights, elevators, house pumps office equipment, other) (______)
equals Total Building Electric Consumption (______)

Fossil Fuel Consumption Summary
(hours · MBH = MBTU)

Outdoor Air Temp.	Hours of Occur.		Hourly MBH	=	Total Con-sumption
92	(____)	·	(____)	=	(____)
87	(____)	·	(____)	=	(____)
82	(____)	·	(____)	=	(____)
77	(____)	·	(____)	=	(____)
⋮	⋮		⋮		⋮
27	(____)	·	(____)	=	(____)

Total HVAC Consumption (______)
plus non-HVAC Consumption (Domestic hot water, food process, other) (______)
equals Total Building Fossil Fuel Consumption (______)

Step A1 Calculate zone cooling requirements

Step A1 requires calculation of minimum required zone cooling capacity.

In air conditioning language, a **"thermal zone" is one or more rooms with similar cooling and heating load patterns** such that, at any one time, they will all need about the same proportion of either cooling or heating. Most buildings have a number of zones, as shown in **figure SP-5.** There is usually at least one zone for each solar exposure. These natural thermal zones should correspond to the HVAC system temperature control zones. As part of the data collection, building areas should be organized into these zones. This will make it easier to calculate loads on a zone-by-zone basis.

Zone peak thermal cooling load is the greatest load likely to occur in that zone at any time during the entire year; it is calculated for the time of day and month of year when that peak occurs. For example, in Atlanta, Ga., where the sample problem building is located, if glass solar loads are large relative to other loads, the east zone will probably peak in the morning in June or July, and the south zone about noon in October or November.

Typical assumptions for a zone peak cooling estimate are:

Glass: clean with blinds or draperies closed.
Atmospheric haze factor: yes, if haze is common at design conditions.
Industrial haze: no, unless this haze is very common at design conditions.
Lighting, equipment, people: greatest amount reasonably expected.
Building thermal storage: yes.

In a zone system response analysis, several items are determined:

How much reheat energy, if any, is to be added to the zone thermal load by the zone temperature control system. Reheat is usually added by a reheat coil or by mixing warm and cold air via a multi-zone or dual-duct arrangement.
The psychrometric processes and coil selections determine air and water flow, along with cooling and heating energy which must be produced by the cooling and heating equipment. This leads to determining the performance of energy consuming equipment.
Distribution of loads between zone equipment and central system equipment.

Zone system response matters most in central systems where reheat is possible or where zone equipment handles only part of the zone load, the rest being handled by the central system.

(The sample problem is an example of both situations. Perimeter spaces are ventilated by a constant flow of cold air from the interior zone central system. This cool air handles part of the exterior zone peak cooling load, thereby reducing the load on the fan-coil units. In cold weather, this cool ventilation air adds to the zone heating requirements. Fan coils must then warm up this air, adding a component of "reheat" to the system).

The load estimate must be arranged so each item of system response may be easily calculated. The zone temperature control of most HVAC systems responds to a zone thermostat, which in turn responds to changes in zone sensible heat load. This sub-total should therefore be calculated separately.

If ventilation air is treated or mixed at the central equipment, the ventilation air load should be calculated later as part of the central system thermal load, since it does not affect the zone thermostat directly. If the system uses a ceiling return plenum and part of the roof load or heat from the lights goes directly into the ceiling plenum, then this portion of the sensible load, called "return air load," should also be kept separate. Return air loads are usually handled by the central equipment not the zone equipment.

Energy-consuming **zone equipment** is required to handle its own loads as described in the system response sub-step. The aim is to:

Find the zone equipment energy use rate at the peak design cooling load condition.
Find the capacity and performance characteristics of each item of zone equipment, so that its energy consumption can be calculated at all other operating conditions.

A decision to make at this point (and whenever equipment performance is evaluated) is whether to use average or specific performance data. If the aim of the analysis is to compare architectural alternatives or different types of HVAC systems, average data for typical equipment will do, since errors in equipment energy consumption will affect both estimates proportionately. If the aim is to evaluate savings available through different equipment, specific performance data for the proposed equipment should be used at calculated percentage load and operating conditions.

If it is a **central system,** the energy consuming zone equipment may consist of a zone fan which circulates zone air through zone cooling and/or heating coils. However, these coils are supplied with cooling or heating energy (chilled water, hot water or steam) from central cooling and heating equipment.

If the system is a **self-contained unitary type,** zone equipment will include the cooling and/or heating equipment. In this case, the equipment may call for input of utility energy from two sources. A gas-fired absorption type cooling unit requires electric power to operate fans and pumps, and natural gas to run the absorption cooling cycle.

Step A2 – Calculate central system cooling requirements

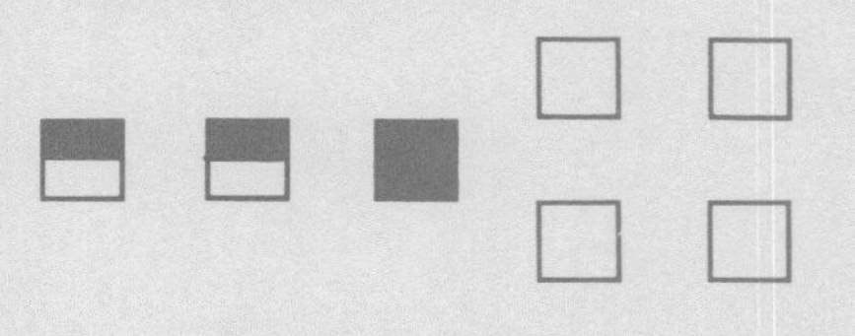

Step A2 is required for central systems only. It compares required cooling loads and actual installed capacity for the central HVAC system, as done earlier for the zones in Step A1. Calculate the central system thermal load, determine the system response, and select each item of equipment.

The **central system thermal load** is the sum of that portion of the zone loads handled by the central system and the central ventilation air load, duct heat gain or loss, and central fan motor heat.

If zone peak loads were to occur simultaneously, the zone load component would be the sum of the central equipment cooling loads from the zone system response sub-steps. In most buildings, individual zones reach peak cooling loads in different months and times of day, and there is a great diversity of people, lighting and equipment loads. So a new "block load" for the month and hour must be computed when the combined cooling load on the entire building is greatest – usually about 4:00 PM in July or August.

If there is a likelihood of reheat in any zone, the load on that zone should be run separately. That way the zone response can be calculated to include reheat in the central system load. If reheat is not involved, the zones can be grouped in a single **"block load,"** as in the sample problem. This is quicker.

There is only one major difference between typical assumptions for the zone peak cooling thermal load and the block peak cooling thermal load. **For the block load, diversity factors should be applied to the lighting, people and equipment loads.** This reflects the fact that even though a zone may encounter zone peak design value at some time, maximum values are unlikely to occur at every zone at peak load time.

If you reduce zone cooling load because part of the heat gain from lights is being returned through a return air ceiling plenum, be sure to include this item in the central system thermal load.

The central system response includes the impact of the central system arrangement, control and sizing.

Fan motor heat should be noted. In a draw-through arrangement, fan heat raises the supply air temperature to the zones. In a blow-through cooling apparatus, however, fan heat raises the temperature entering the cooling coil, but does not change the supply air temperature to the zones.

The performance of cooling and heating coils is part of the central system response. That is because these items do not consume energy directly, but do establish the air quantities and water flows for the energy consuming equipment. The effects of "outside air exhaust air" heat exchangers and "glycol run-around coil systems" are also considered in system response because they reduce the load to the energy consuming heating and cooling equipment.

The output of the system response consists of the cooling and heating energy plus air and water flows for each item of central equipment.

The output due to the optimum environmental requirements are compared with installed capacities to determine if any equipment has an unreasonable amount of excess capacity.

Step A2 is completed by using the above maximum capacity requirements and the equipment performance ratings to determine the utility energy input to each item at central equipment. These are added up to obtain central equipment peak demand. As in Step A1, a fossil fuel energy source will also be required if the refrigeration equipment is of the absorption type.

Step A3 Calculate zone heating requirements

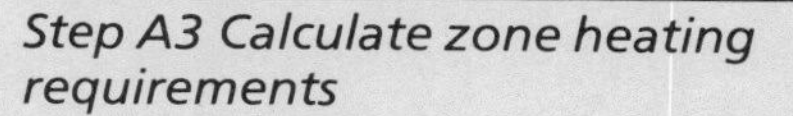

Step A3 repeats step A1 except for the maximum heating situation. This usually occurs at the beginning of the occupied period, before the sun and internal loads from lights and people have begun to reduce heating loads. Morning warm-up loads should be included if night setback is used to lower room temperature during the unoccupied period.

Ventilation air tends to be a large part of the heating load. The energy planner should establish whether the ventilation air dampers must remain fully open at peak heating load, or whether they can be closed (with exhaust fans off) during morning warm-up until heat gains from sun, people and lights reduce the zone heating loads.

In most climates, **infiltration** is much greater in winter than in summer, and may also be a large component of the heating load. This is especially true if the excess of ventilation air quantity over exhaust air quantity is reduced so there is little "pressurization" of the building.

Excess heating capacity generally hurts efficient operation less than does excess cooling capacity. What is more, it is useful in warming up the building quickly in the morning if night set-back was used. For this reason, most designers still include a safety or **"morning pick-up"** factor in their heating design calculations. In any heavy building, the possible advantages of night setback energy-savings as against greater equipment size for morning pick-up should be looked into for their effect on initial cost of equipment, on morning winter peak demand and on operating costs.

Step A4 – Calculate central system heating requirements

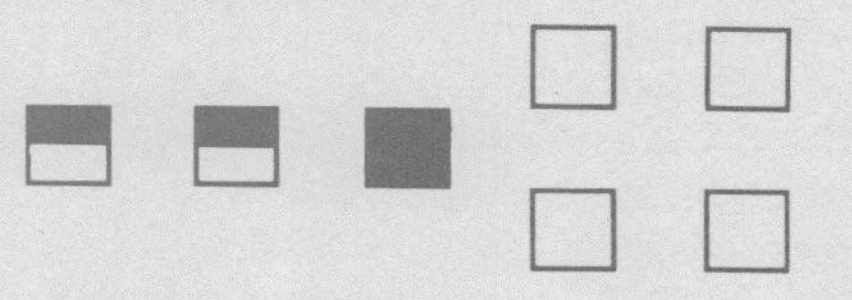

Step A4 repeats step A2 except for the maximum heating situation. The heating design uses the same approach as for the cooling design, except that the block load for selecting central equipment is usually taken as the sum of the zone peaks. This is valid because design heating loads are usually calculated without taking credit for sun, lights or people. **Also, all zones are assumed to require full heating at the same time.**

The comments under Step A3 for zone heating equipment regarding ventilation, infiltration and safety factors apply equally to Step A4.

Part B – Calculate HVAC peak energy demand

Fossil fuel energy costs do not usually include demand charges. That means usually that only electric energy peak demand needs to be computed.

If monthly demand charge is likely to be set by one month's peak, this information is available from Part A (the design capacity verification steps). Usually, a monthly demand will be required, however, at least for electric energy. The easiest way is to **prepare a demand load profile.** Weather data may then be used to find the highest (or lowest) temperatures in each month. The monthly peak demand is read directly from the profile.

Most heating systems lack the demand limit controls required to warm up a building gradually if night set-back is used. If the building cools off over a 50°F weekend, the heating will operate at full capacity on start-up until the building reaches normal operating temperature. After a 20°F weekend, it will operate at the same demand rate, but take longer to warm up the building. The point is that **heating demand will probably be established by capacity of the heating equipment,** not by outdoor temperature – unless a sophisticated demand limit control is used.

If the HVAC system uses fossil fuel for heating and electricity for refrigeration, and if the month's maximum temperature is high enough to require refrigeration, then maximum electric demand will usually come at the highest monthly outside temperature. If refrigeration is not required that month, or if it is non-electric, the greater use of electricity for lighting or elevators in early morning hours may result in a peak electric demand on a cold morning. Natural lighting (daylighting) makes this even more likely.

If electric heat is used, peak demand may come at the coldest monthly temperatures during morning warm-up. Night set-back, off-peak electric hot water storage and automatic demand limit control systems may change the demand characteristics.

If one cannot predict whether peak demand will come during cooling or heating, both cooling and heating utility energy profiles must be plotted and the higher value for each month selected.

The **cooling demand profile** is plotted by using utility energy demand from Steps A1 and A2 to obtain the maximum design cooling point, and by calculating one or more points in Step B1 for the other point(s), **(figure 3-16 – Peak demand profiles, Steps B1, B2 and B3).**

The **heating demand profile** is plotted in the same way, by using demand values from Steps A3 and/or A4 for the maximum design heating point and for one or more points found in Step B2.

Monthly peak demand is found in Step B3 by entering the peak cooling profile at the highest monthly temperature and the heating profile at the lowest, to obtain peak demand KW. **The higher of the two demand values is the monthly peak KW.**

Step B1 – Calculate and plot profile points – cooling demand

Step B1 calculates and plots utility energy demand at one or more points besides the point already calculated in Steps A1 and/or A2. If peak demand results from refrigeration load (usual with electric powered refrigeration equipment), the outside temperature range of the profile should cover the range of highest monthly outside temperatures during which refrigeration is required. Two profile points, one near the upper limit and one near the lower, are usually enough if there is no enthalpy control type economizer cycle. An extra point at the average changeover temperature is needed for enthalpy control. **(Figure 3-17 – Typical refrigeration system demand profiles)**

Thermal load is calculated for each profile point temperature in a manner similar to Steps A1 and/or A2.

If the profile point temperature is low enough to require heating in any zone, a single "block load" estimate cannot be used for a multiple-zone building. If it were the cooling load in an interior zone or a sunny exterior zone would cancel some or all of the heating load, and end up with a low estimate for both cooling and heating loads. The sun shining in to a south office does not add heat to a north office with most system designs; it creates a cooling load on the south office and the north office needs just as much heat. The only precise way to estimate loads at cold temperatures is to calculate each zone separately and to keep a separate accounting of cooling and heating energy.

Outside dry bulb temperature for the thermal load estimate is the profile point temperature. Outside wet bulb temperature for cooling profiles should be the highest "coincident wet bulb" from the weather data. If the Air Force Weather Data is used, **(figures SP-7a & 7b)** choose a month with a maximum temperature close to the profile point, and use the coincident wet bulb temperature corresponding to the profile point bin.

Solar load is found for the specific month selected, and for the time of day when the combination of solar, transmission, internal and ventilation air loads is greatest. There is generally less solar haze (water vapor in the air) in winter because the moisture content of the outside air is lower.

Equivalent temperature differences for walls and roof should be corrected for the lower outside temperature.

Numbers of **people** and diversity factor applied to **lights** and **equipment loads** should continue to reflect the "greatest reasonable load" philosophy rather than the "greatest possible load."

The system response step is as in Part A before, except that reheat is more likely at conditions less than full capacity due to cooling loads in some zones being very low. Heating may even be required. **Refer to the "system response" discussion in the appendix** before trying to calculate system response.

Consumption calculations for equipment energy are also as described earlier, except that **operating conditions differ.** For example, at the peak design cooling point, the refrigeration machine may have been operating at full load, with 85°F condenser water and 44°F chilled water and a power input of 0.75 KW per ton. At a lower profile point, it may be operating with 70°F condenser water at 40% load and a power input of 0.68 KW per ton.

Moreover, some equipment may not be operating continuously at this lower outside temperature. The cooling tower or air-cooled condenser fans, the equipment room and elevator room exhaust fans all fall in this category.

The cooling energy demand profile can consist of the entire HVAC demand, any group of components such as the

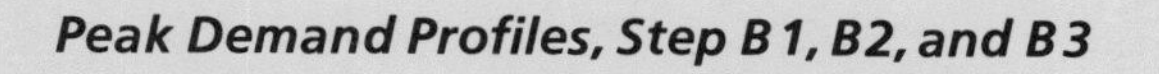

Peak Demand Profiles, Step B1, B2, and B3

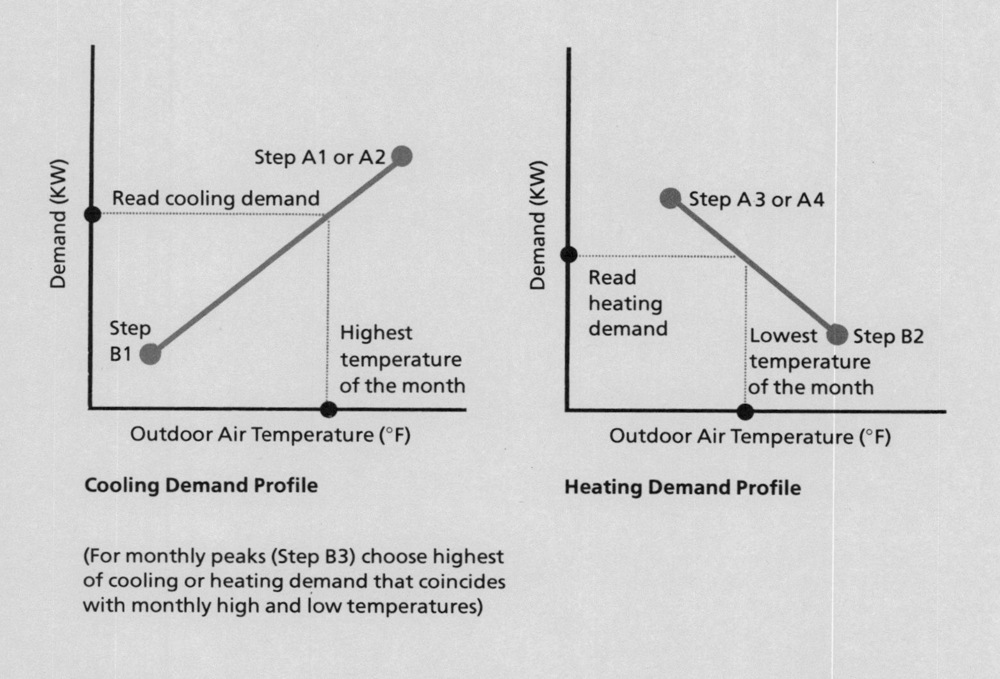

(For monthly peaks (Step B3) choose highest of cooling or heating demand that coincides with monthly high and low temperatures)

Figure 3-16

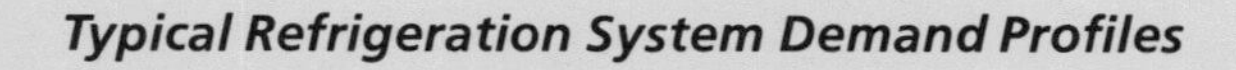

Typical Refrigeration System Demand Profiles

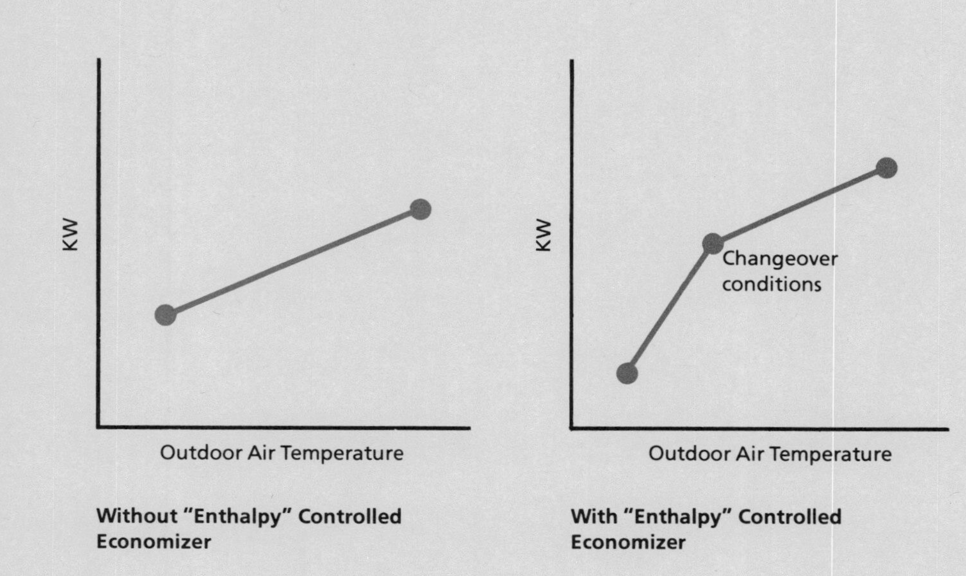

Figure 3-17

refrigeration cycle (refrigeration machine, chilled and condenser water pumps and cooling tower fans), or the refrigeration machine alone. Any items not included in this profile must be added separately in Step B3 to obtain total HVAC monthly peak demand.

Step B2 – Calculate and plot profile points – heating demand

Step B2 calculates and plots utility energy demand at the additional profile point(s) for the heating demand profile. This step is commonly not required for electric energy demand unless the building uses electric heat. It may be required, however, for a different heating fuel if cost of that fuel includes a monthly demand charge.

The peak heating design point was found from Step A3 or A4. At least one other point is needed, and if operation of the system is non-linear, another point is needed at each change. For example, if there is heating with a heat pump which can handle the entire heating load down to 35°F degrees outside, but electric strip heat must be used to supplement the heat pump below 35°F degrees, a profile point is needed at 35°F as well.

If the outside temperature at which peak heating load is zero can be easily determined, heating load calculations are not required and energy consumption consists simply of operating fans and pumps.

Thermal heating load calculations are similar to those in Steps A3 and A4. If humidification is involved, coincident wet bulb temperature is the lowest to be expected at the profile point bin.

System response and equipment energy consumption calculations are as before. Similarly, plotting of the heating demand profile follows the procedure used in Step B1.

Step B3 – Monthly peak demand summary

If the utility bases its demand charge on actual monthly demand, or if no one knows which month's peak energy demand will set the demand charge, the peak HVAC energy demand must be computed for each month as is done in the sample problem. Separate profiles will be required for each purchased energy source. For a month by month estimate, first, obtain each month's highest and lowest temperatures from the weather data. The monthly peak demand that corresponds to those temperatures can be easily read from the cooling or heating demand profiles. **If it is not known whether heating or cooling demand will be greater for any month, simply read both and use the highest,** as shown previously in **figure 3-16.**

Part C – Calculate HVAC average energy consumption

The "modified bin method" is used to illustrate the procedure for calculating utility energy consumption. It calls for preparation of a profile of average energy consumption at the various outside temperatures.

The energy planner should consider two operating modes for the HVAC system. The first, the normal **"occupied" period,** occurs when the HVAC system is satisfying normal operating conditions; the second, the **"unoccupied" or standby heating period,** occurs during nights and weekends. That is when cooling equipment is usually shut off and heating systems run at reduced capacity to prevent the building from becoming too cold.

To account for these two operation modes, the energy consumption estimate must be done in two separate subestimates, one for the occupied periods and one for unoccupied periods – each with its own energy consumption profiles and energy summaries. If the rate schedule requires monthly energy summaries then more effort will be required than if annual summaries are required. Total building HVAC energy consumption is obtained by adding the consumption for both periods.

The calculation for each operating mode is similar

- Select the profile points that coincide with the operating conditions.
- Calculate energy consumption of each item of equipment for the average cooling and/or heating loads at the selected profile point.
- Plot these values from each profile point to form the average energy consumption profile.
- Obtain average hourly energy consumption for each outdoor air temperature bin from the profile.
- Obtain monthly hours of occurrence of each outdoor air temperature bin from the weather data (or annual hours for a year's total).
- Total the monthly (or annual) average HVAC energy consumption by multiplying average hourly energy consumption by monthly (or annual) hours of occurrence for each outdoor air temperature bin and adding them up.

If HVAC equipment uses more than one energy source, individual profiles and monthly energy summaries are required for each source whose consumption varies with outside temperature.

If separate energy estimates are required for each item of equipment, it is easier to use the **equivalent full-load hours** method for equipment whose energy consumption is determined solely by hours of operation.

Step C1 – Calculate and plot profile points – average energy consumption during occupied hours

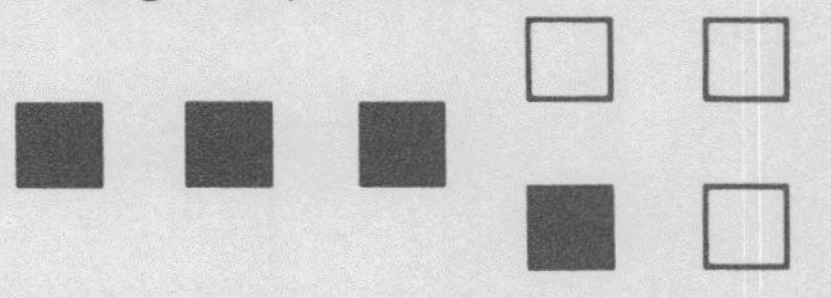

In Step C1, average consumption of each utility energy source is calculated and plotted at each profile point for the occupied period. **The procedure is similar to that used in preparing the demand load profiles. It differs**

in that the objective is changed to predict the hourly HVAC load likely to occur at the selected outdoor air temperature, averaged over the entire system operating period, rather than at the highest load under selected extreme conditions.

Since the aim is the average hourly HVAC load at the profile point temperature, one must average over the HVAC operating period those HVAC loads that are due to factors which change independently of outdoor air temperature (solar heat gain, and internal heat gain from artificial lighting, equipment and people). If the HVAC system operates on a 24-hour schedule, but the building is mostly unoccupied at night, one can increase accuracy by making separate analyses for the two different conditions.

A profile point should be selected near the upper range of summer temperatures and another near the lower range of winter temperatures. Additional points should be chosen at those outside temperatures where the slope of the profile could change, as follows:

About 70°F if an enthalpy type of economizer cycle is used.
About 55°F, or the point below which the refrigeration machine will not be operated.
The temperature below which heating will be required.

After selecting profile points, weather data must be organized to obtain the average coincident wet bulb temperature and solar condition at each point. The relationship between outside temperature, month of year and average amount or percent of sunshine is set by plotting average daytime temperatures for the various months, as shown in **figure SP-29 – Solar relationship to outside temperature.** From the curve, a month (with coincident percent sunshine for the average load profile) may be chosen for calculating solar loads at each selected outdoor air temperature.

The most appropriate type of thermal load estimate for each point must also be selected. The objective is to insure that the averaging process does not overlook reheat or underestimate loads.

There are several types of calculations from which to choose:

The "block load" format: thermal loads and system response of all zones combined into a single estimate.
The "zone-by-zone" format: separate estimates are prepared for each zone individually, but all hours of the occupied period are combined into a single zone estimate.
The "zone-by-zone, hour-by-hour" format: thermal load and system response are calculated separately for each hour of each zone.

The block load format is correct only if one type of energy (heating or cooling) is required at any operating condition at that profile point. For example, if it is 90°F outside and all zones require cooling regardless of exposure or time of day, the block load format is suitable and is quickest.

The need for zone or hourly calculations is based on two principles:

A cooling load in one zone cannot offset a heating load in another. If a north zone which needed 20,000 BTUH of heat on a 50°F day were combined in the same block load with a sunny south zone which needed 20,000 BTUH of cooling, the average block load would be zero. Yet the real load is 20,000 BTUH of cooling plus 20,000 BTUH of heating. Hence the zone method is much more accurate.
Within each zone, a cooling load at one hour cannot offset a heating load at another hour. If a west office on a cool day requires 20,000 BTU of heating in the morning and 20,000 BTU of cooling in the afternoon, the true average hourly load for the entire day is not zero. If the system runs 10 hours per day, the true average load is 2,000 BTUH of cooling plus 2,000 BTUH of heating. In this situation, zone loads must be computed hour-by-hour to keep separate track of cooling and heating requirements.

In a manual calculation, the time required to do the calculations expands considerably by going to zone or hourly calculations, and this calls for good judgment. If a purpose of the analysis is to compare two types of HVAC systems with different system responses, the accuracy of the better techniques is justified. Otherwise the improved accuracy is probably not justified. With computer-aided calculations, the zone format should be used with hourly figuring of loads.

With the block load method, solar load is calculated separately for each exposure, as shown in **figure 3-18 – Average glass solar load: block load format.**

The **shading coefficient** is for an average combination of glass and shading device (including overhangs and reveals). The haze and dirty glass factor also is for the average condition (0.80 to 0.90 is typical).

The **average solar factor for each exposure** is the sum of the solar heat gain per square foot for each hour of the day, divided by hours of HVAC system operation. If the operation is a school which does not operate after 3:00 PM solar loads may be excluded after the system shuts down. Even so, some of this this heat may remain stored in the building and become a "pull-down" load the following morning.

The **sunshine factor** is the decimal equivalent of average percent sunshine for the month of estimate.

The **temperature difference** for glass and wall transmission should be the actual "outside-to-inside" difference. With the block load method, it is impractical to try to consider heat storage and time lag as they affect building transmission load. That is because one does not know whether outside temperature is rising or falling at this "average" condition.

As roofs tend to be exposed to solar radiation for an entire day, one should add about 10 degrees to the **actual temperature difference** when calculating roof transmission load. This makes for a more realistic average equivalent temperature **(figure 3-19 – Average heat transmission loads).**

Loads from lights, people and equipment should include three corrections, as shown in **figure 3-20 – Average lighting loads.**

The **average diversity factor** reduces the peak load to account for the fact that some lights and equipment will not be operating and some people will be absent. For instance, the installed lighting wattage may be 50 KW, but during the normal occupancy from 8:00 AM to 6:00 PM, only an average of 40 KW may be in use, for a 0.8 diversity factor.

The **hours factor** corrects the load to an average load during system operation. If the system ran from 6:00 AM to 10:00 PM, the hours factor for the lights would be the actual 10 hours of lighting load divided by 16 hours of HVAC system operation, to give the average hourly lighting load on the HVAC system.

The **plenum factor** reduces the load charged to the zone if lights are installed in a ceiling used as a return air plenum. The remaining load is charged as return air heat gain to the central apparatus.

Ventilation air quantity should likewise be corrected to obtain an average cfm during HVAC system operating hours. This is important if the ventilation air damper is closed during morning pull-down building occupancy or after normal operating hours, or if the HVAC system is a variable air volume type where ventilation air is often a fixed percentage of the variable supply air quantity. Closing the outside air damper does not help much if building exhaust fans are kept on **(figure 3-21 – Average ventilation and infiltration loads.)**

The infiltration load is similar to the ventilation air load, except for two items. First, the **hours factor** may be inversely related to the ventilation air time period. That comes about if the building is sufficiently airtight and/or the ventilation air quantity is large enough to prevent infiltration by pressurizing the building. Second, the **diversity factor** should be related to average weather condition at the profile point. Wind velocity and outside temperature (stack effect) affect infiltration. In the Atlanta, Ga. climate, for instance, most designers neglect infiltration in summer, but allow for it in winter, based on type of windows, wall construction, outside entrances and building height.

If the zone-by-zone estimate is used, individual zone solar, transmission and internal loads are first calculated and summed separately for each zone. This way the zone system response can also be calculated separately. It prevents an interior zone cooling load from offsetting a perimeter zone heating load at a cold outside temperature profile point.

If the zone-by-zone, hour-by-hour estimate is used, the format changes. Individual zone solar, transmission and internal loads are figured separately for each hour of system operation, as shown in **figure 3-22 – Hour by hour zone thermal loads (MBH).**

Outside temperature remains the same for all hours since average hourly load is being computed at a specific profile point (outside temperature). Transmission load is, therefore, the same for all hours.

Solar load is calculated separately for each hour. Glass area, shade coefficient, haze factors and percent sunshine are usually combined into a constant multiplier used with the appropriate hourly solar heat gain value. In case of architectural reveals, fins or overhangs, it is more accurate to recalculate the shade factor on an hourly basis.

The hour-by-hour format allows the average hourly load to reflect the typical operating schedules for lights, equipment and people, as shown in **figure 3-22.**

Infiltration air quantity may vary with each hour to reflect the effect of pressurizing the building with ventilation air.

The **ventilation air load** should be treated as a zone load, if air is supplied directly to the zone equipment. If ventilation air is pre-conditioned at the central system, it should be figured as a central load.

Whenever zone **latent load** is handled by zone dehumidifying coils, it should be calculated as a zone load. If **dehumidification and humidification** are handled by central equipment, the latent load from all zones should be figured as a single block load.

It is usually accurate enough to calculate either zone or block latent loads as a single average daily load. If both humidification and dehumidification are closely controlled and the latent load reverses during the day, an hour-by-hour calculation is called for. It is done the same way as the zone sensible load estimate.

Throughout the figures in Chapter 3 certain data is printed in green. This indicates data gathered, calculated or derived by the energy planner during the course of his services.

Average Glass Solar Load: Block Load Format

Exposure	Glass area		Shading coefficient		Haze factor Dirt factor		Average solar factor		Sunshine factor		Heat gain (BTU)
North	20,818	×	.53	×	.9	×	11	×	.65	=	71,001
East	13,566	×	.53	×	.9	×	55	×	.65	=	71,000
South	17,290	×	.53	×	.9	×	28		.65	=	150,101
West	13,566	×	.53	×	.9	×	55	×	.65	=	231,337

Figure 3-18

Average Heat Transmission Loads

Item	Area		"U"		Temperature difference		Load (BTU)
Glass	65,240	×	1.10	×	(92-75)	=	1,219,988
Wall	50,728	×	.16	×	(92-75)	=	137,980
Roof	13,044	×	.10	×	(102-75)	=	35,219

Figure 3-19

Average Lighting Loads

Installed KW		Average diversity		Hours factor		Plenum factor		Constant		Heat gain (BTUH)
429	×	.8	×	10/16	×	.75	×	3413	=	549,066

Figure 3-20

Average Ventilation and Infiltration Loads

Ventilation or infiltration CFM		Average diversity		Hours factor		Temperature difference		Constant		Load (BTUH)*
35,000	×	.7	×	9/16	×	(92-75)	×	1.09	=	255,367

*Sensible heat – the latent heat calculation is similar, but uses 0.68 for the constant.

Figure 3-21

Hour-by-hour Zone Thermal Loads (MBH)

Hour	Transmission	Infiltration	Lights	Equipment	People	Solar	Total Zone Sensible Heat (MBH)
6	– 360	– 40	0	0	0	0	– 400
7	– 360	– 40	0	0	0	57	– 343
8	– 360	– 40	176	0	39	115	– 70
9	– 360	0	176	10	39	383	+ 248
10	– 360	0	176	10	39	523	+ 388

Figure 3-22

The zone thermal load and zone system response calculations must follow the same format. When using the hour-by-hour, zone-by-zone estimate, zone loads are figured on an hourly basis, and the various hours are averaged after the system response is calculated. Central system response is calculated on a block load basis, using average hourly zone and reheat loads. **Figure 3-23 — Hour-by-hour zone system response (MBH)** shows how an hour-by-hour, zone system response calculation would look for a four-pipe sequenced cooling and heating control fan-coil system at a cold profile point. Assumed was a central ventilation system, and infiltration loads are included in the zone sensible heat calculations.

Equipment performance is figured to determine utility energy consumption of each item of equipment at each profile point.

The average energy consumption profile for the occupied period is plotted using energy consumption va ues calculated for each profile point. A profile is required for each purchased source. **The profile may include only equipment whose consumption varies with outside temperature,** such as refrigeration compressors and heating boiler; or could include some or all other equipment using that energy source. To keep a separate accounting of energy used by a particular item of equipment, record the item separately throughout the summary process.

Step C2 – Calculate and plot profile points – average energy consumption during unoccupied hours

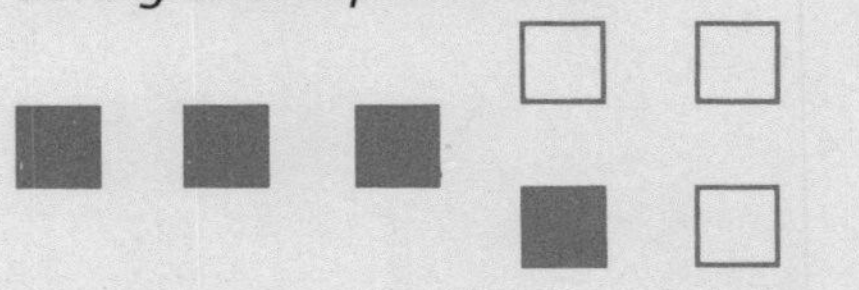

Step C2 calculates and plots energy consumption at each profile point during the unoccupied period.

The energy involved is usually that required to heat the building when unoccupied on nights and weekends. Cooling equipment is usually off. If cooling is required for security guards or critical areas such as computer terminals, this cooling energy must be figured. But consider providing a separate small cooling system or some way to cool these areas without inefficiently operating the large central system.

Hour-by-hour Zone System Response (MBH)

Hour	Zone SH Load	Fan Coil Motor Heat	Vent Air	Adjusted Zone SH	Cooling Coil Load (MBH)	Heating Coil Load (MBH)
6	− 400	166	0	− 234	0	234
7	− 343	166	0	− 177	0	177
8	− 70	166	− 188	− 92	0	92
9	+ 248	166	− 188	226	226	0
10	+ 388	166	− 188	366	366	0

Figure 3-23

If cooling is not available during the unoccupied period, a simple approach can be used. Only two profile points are needed: one near the outside design heating temperature, the other at or near the outside temperature at which standby heating is first required. Since there is no cooling, the block load may be used instead of the hour-by-hour method. The only inaccuracy that could come up is if the heating system is so poorly zoned or controlled that one area is much warmer than intended. This creates a greater temperature difference between outside and inside air than estimated, and hence greater heat transmission through the building skin. The calculation for each profile point usually follows the procedure shown in the **Master Chart (figure 3-15).**

Most of the unoccupied period is at night, so calculations usually neglect any credit for solar heat gain. If window areas are large, and a lot of solar heat gain is received during the day on Saturday and Sunday, deduct this as a credit from the heating load.

In heating calculations, power input to all fan and pump motors (less power lost to the equipment room due to motor inefficiency) is in the end converted to heat energy and is a credit against the heating load. The same is true for corridor lights which are on all the time.

Ventilation air dampers are usually closed during an unoccupied heating period. If the air distribution system must be turned on to provide heating, leakage may occur through the dampers, even when fully closed.

When the building is empty and the air distribution system is off, more infiltration will occur than during the occupied period, because the ventilation system no longer pressurizes the building.

If night setback of room temperatures is used, calculations should be based on the setback temperature. Building temperature does not instantly drop to the setback temperature. This correction can be made by adding from 2°F to 5°F to the night setback room temperature. For greater precision a separate calculation can be done to estimate temperature drift, and the hour-by-hour estimate format can be used.

If the simple estimating format is used for night setback, the morning warm-up load is about equal to the night cool-down credit, so there is no need to figure this for energy consumption.

Thermal load calculations can usually take advantage of a block load format, as shown in **figure 3-24 – Block load heating estimate.**

The central system response determines which items of equipment must be in operation, as well as the heating energy load. Equipment energy consumption calculations can then determine consumption for each item.

Energy consumption for unoccupied periods may be plotted as a total system profile or as individual profiles of variable response equipment, in the same way as for the occupied period in Step C1.

Block Load Heating Estimate

Item	Sq. Ft. or CFM		"U" or constant		Temperature difference		BTUH
Glass	65,240	×	1.1	×	(65 – 10)	=	3,947,020
Wall	50,728	×	.16	×	(65 – 10)	=	446,406
Roof	13,044	×	.10	×	(65 – 10)	=	71,742
Infiltration	8,435	×	1.09	×	(65 – 10)	=	505,678
	Total building heating load					=	**4,970,846**

Figure 3-24

Step C3 – Prepare monthly (or annual) HVAC energy consumption summary

Step C3 is the HVAC energy consumption summary, as shown in the master chart **(figure 3-15).** If individual monthly summaries are not required, a single annual summary is obviously easier. Monthly energy summaries are required in any one of the following conditions:

- A monthly demand charge is involved.
- The utility energy cost per unit depends on monthly usage.
- Estimates are to be compared with previous monthly bills for an existing building.

Bin method summaries were described earlier. The data required is 1) how many hours occur in each temperature bin during the occupied period and 2) an occupied period average energy profile. Taking each temperature bin one at a time, find the energy consumption rate from the profile and multiply it by the number of hours to obtain a total energy usage for that bin. After calculating all the bins, add them up for the monthly (or annual) total.

A separate summary must be made for each energy type and for each item of equipment to be tracked separately.

If separate equipment totals are kept, it is best to figure consumption of uniform response equipment by adding total hours and multiplying by the average energy consumption rate. The bin method is not required unless energy consumption varies with outside temperature. Occupied and unoccupied period energy estimates are combined for the monthly total for each item of equipment. These monthly equipment totals are combined to obtain HVAC monthly utility energy consumption.

Estimate non-HVAC energy use

Electric power for lights, elevators and other equipment usually comes through the same electric meter as electricity for the HVAC system. The energy estimate must therefore be arranged so as to permit adding all items to yield the total. Two principles should be observed in calculating energy for this equipment:

- The effort for greater estimating accuracy on this item should not exceed its significance in the building's total energy consumption.
- Greater accuracy and detail will be needed should a modification involve this item.

Elevator equipment is a good example. If there is no interest in modifications, a simple allowance from 50 to 100 KWH per day per elevator is appropriate. The elevator manufacturer could provide a closer estimate. If modifications are being considered, the manufacturer is best equipped to assess their effect on his equipment.

Estimating non-HVAC energy use involves the same three parts as the HVAC energy estimate.

A Compare equipment capacity with requirements.
B Calculate peak energy demand.
C Calculate average energy consumption.

Part A. Compare equipment capacity with requirements

This is especially important for the lighting estimate because artificial lighting is commonly a large segment of total building electricity consumption. Also, heat gain from lights is a major part of the HVAC system cooling load. Reduced lighting power input will usually permit a corresponding input reduction in HVAC system fan and refrigeration system power.

With the HVAC estimate, the design capacity/requirement comparison was best accomplished by recalculating HVAC loads independently of any available original design information. The same approach should be used for the lighting. Actual requirements should be determined methodically by investigating:

- What activity really takes place in this space.
- How much artificial illumination is really required.
- How artificial illumination can be provided most efficiently.
- Whether this illumination is needed constantly. If not, how it can be most effectively turned off when not needed.

In an existing building with multiple elevators, actual requirements may often be determined simply by monitoring elevator operation and modifying the operating program by trial and error. The best operating program is one that provides lowest energy consumption with acceptable waiting times.

Part B – Calculate non-HVAC peak energy demand

When the same electric meter serves lighting, elevators, HVAC and other equipment, and a demand charge is involved, peak demand for each item must be estimated at the peak hour and month of total building demand. It does not matter if one item reaches a higher value at a different time, so long as this value in combination with the other items does not raise the total building peak.

Peak demand values for lighting are normally obtained by taking an inventory or estimating lights likely to be in use at time of peak demand. This total is reduced by a diversity factor to account for lights which are burned out, or located in areas that are out of service for renovation or change of tenants. Demand diversity factors of 10% to 15% are common.

Part C – Calculate non-HVAC average energy consumption

The two types of manual energy consumption calculations that apply to all but HVAC equipment were discussed earlier. They are:

Constant rate operation. Whenever the item is operating it consumes power at the same rate. The energy planner needs to find how many hours per month it actually operates. Sometimes this is done by taking elapsed time in which the item is available (the building occupied period, for example) and multiplying this by a percent usage factor to obtain actual operating hours.

Variable rate operation. The item operates continuously but power consumption varies because the load on the equipment for some reason changes. An elevator is typical of a modulating load. The motor-generator (MG) set or hydraulic pump runs all the time, but the load changes as the elevator goes up, down or stops at a floor. An elevator may be a combination of both on-off and variable, if the MG set shuts off during non-use.

Energy consumption of variable input equipment is figured by using the equivalent full-load hours method. The procedure is as follows:

Estimate total hours of operation per month.

Estimate average percent power input during the entire period of use.

Item 1 times Item 2 equals the "Equivalent full-load operating hours" per month.

Equivalent full-load operating hours per month times energy input at full load equals monthly energy use.

The bin method does not apply unless the rate of energy use can be directly related to outside temperature.

Special Profile: Effect of Lighting Reduction on HVAC

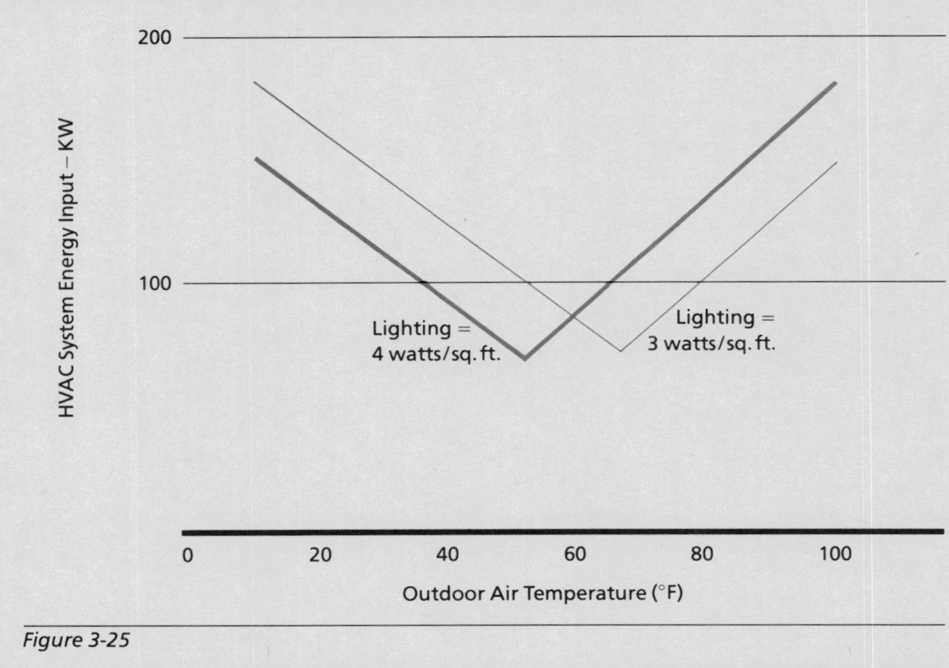

Figure 3-25

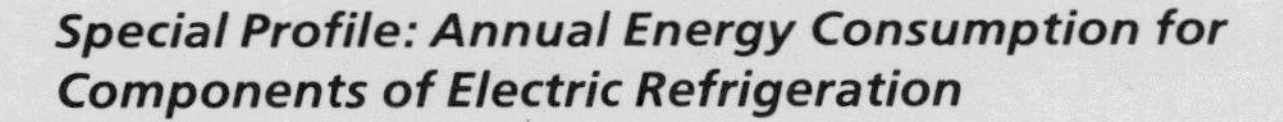

Special Profile: Annual Energy Consumption for Components of Electric Refrigeration

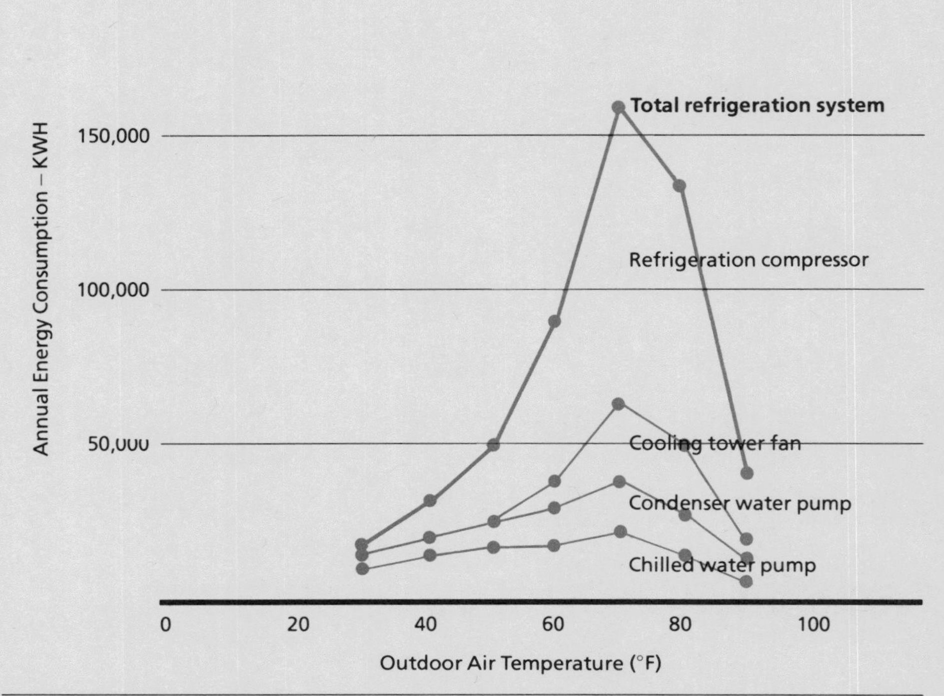

Figure 3-26

If HVAC system energy consumption is being estimated using a computerized hour-by-hour method, a profile of percent power input must be developed at each hour of the day for lights and all building equipment. The computer uses this profile to calculate the energy consumption of equipment. It also calculates the load on the HVAC equipment resulting from this energy consumption within the air-conditioned space. Most computer programs have standard equipment operating schedules from which the energy analyst can select the most appropriate one.

Total HVAC and non-HVAC energy use

The building's total month and annual peak energy demand and average energy consumption is then obtained by adding the contributions of the calculated HVAC, the non-HVAC energy peak demand and average consumption (Parts B and C of non-HVAC calculations).

Preparing additional profiles

A profile of energy use illustrates the relationship between energy use and outside air temperature. It is usually plotted as a graph. This has the advantage of making it easier to remember and/or identify opportunities to reduce energy use.

The profile of average energy consumption versus outside temperature is the only type required for the "bin" method. Other types may be useful in preparing or explaining the analysis, and the bin method is especially suited to preparing additional profiles, since it requires all calculations to be organized around the profile points.

Additional profiles are in order if they can:

Show the effect of a proposed modification on the equipment load.

Show the relative size of components within a total.

Figure 3-25 – Special profile: effect of lighting reduction on HVAC shows the first type. Its purpose is to show the effect of a reduction in lighting power input on HVAC system energy consumption in an all-electric building. When outside temperature exceeds 50°F, building spaces generally require cooling. The reduced lighting load led to a lower refrigeration load and reduced HVAC energy consumption. Below this point, the building required heating, and reduced building internal heat gain from lights tended to increase the HVAC system's heating requirements.

Figure 3-26 – Special profile: annual energy consumption for components of electric refrigeration shows the relative magnitude of energy consuming items in an electrical refrigeration system.

Either the rate of usage, as in the first example, or the total annual usage, as in the second, may be plotted. The rate of usage is of greater interest when discussing demand charges; total energy consumption is more pertinent in its effect on overall energy cost.

Estimating energy costs

The energy estimate's consumption and demand figures must be converted into dollar costs. The results should already be in a format that corresponds to the electricity or fuel supplier's billing procedure. Annual, monthly, or time-of-day figures should have been provided as required.

The current trend, especially with electric companies, is towards more complicated rate schedules. Each company has its own. Furthermore, the energy content of fuel varies with locality and source, and the price of fuel is constantly changing. As a result, the first, most important step in calculating energy costs for the energy estimate is to contact the building's electricity or fuel supplier and determine present and projected fuel costs and the varying energy contents of specific fuels. Suppliers are the only ones who have the exact information and may even do the calculations for you.

The simplest rate structures are based solely on quantity delivered. Most fuel companies, such as oil or coal suppliers, will charge a flat rate per gallon or ton. Natural gas companies bill according to the volume of gas delivered (in dollars/hundred cubic feet) or by actual amount of energy supplied (dollars/therms).

The total bill may include additional charges. This is particularly true in the case of electricity, but may apply to fuel bills, too, especially natural gas. Additional charges may include:

Demand charges. These are added to the customer's bill so the utility company can recover the extra expense of providing generation, transmission and distribution equipment to meet its customer's peak energy requirements. A customer's demand refers to the highest rate of energy consumption during the billing period. Usually a demand meter will record the building's highest energy consumption per 15-minute or 30-minute interval during the billing period. Most electric companies have demand charges and the usual billing unit is the kilowatt or the kilovoltampere. If actual demand is not recorded by a meter, demand charges may be based on the size or the nameplate rating of the customer's installed equipment.

In these cases, the unit used for electric demand is the horsepower.

Sometimes demand charges are based on an "equivalent hours" use of the customer's peak demand. The units are hours obtained by dividing electric consumption in kilowatt-hours by peak demand in kilowatts. For example, if consumption was 10,000 KWH and peak demand was 200 KW, the equivalent hours of peak demand would be 10,000 KWH/200 KW = 50 equivalent hours. Most electric companies use what is known as an "integrated" demand charge, based on the highest amount of energy consumed each month during a specific time interval, usually 15 or 30 minutes. "Integrated" demand charges do not penalize the customer greatly for sudden surges of electric energy use, when large electric motors are turned on. If the surge only lasts for a few seconds or minutes, it will not raise total consumption much during the 15 or 30 minutes, thus keeping the demand charge unchanged.

Some companies provide for customers with highly fluctuating rates by reducing the time interval over which demand is measured to 5 minutes.

For example, if the building's equipment consumed 500 KWH during its peak 30 minute period, the "integrated" demand would be 500 KWH/0.5 hour = 1000 KW.

If during those 30 minutes there was an additional surge of 2000 KW for 3 minutes, the "integrated" demand would be 1200 KW

Additional consumption equals
2000 KW × 0.05 hr. = 100 KWH

Demand equals
(500 + 100) KWH/0.5 hr. = 1200 KW

If a 5-minute interval had been used instead of 30 minutes, the demand with this additional surge would be 500 KWH/0.08333 hr. = 6,000 KW.

Fuel adjustment charges are used to pass on to the customer the difference between actual fuel cost to the utility and fuel cost the utility has included in its basic rate. Costs can change drastically from month to month. Fuel adjustment is usually billed as a flat rate per unit of fuel or energy consumed (such as 1.056 cents per KWH). They too may change drastically from month to month.

Penalties are charged by electric companies to customers with **poor power factors.** Typically, a power factor less than 90% will incur a penalty. Special metering is involved. Generally, larger customers, such as those having a monthly demand greater than 1000 KW, are affected, although the electric company may reserve the option to install metering equipment in smaller buildings.

Power factor penalties are also referred to as Excess Reactive Demand, measured in kilovars (KVAR). Excess Reactive Demand is a vector quantity which relates induction and resistance power supplied to actual resistance power used. For instance, the rate structure may include a charge of $0.20 per excess KVAR. Some companies will give the customer credit for a very good power factor.

Consumer or service connections charges are usually a flat rate per month billed to customers for all costs involved in hooking up, keeping accounts and maintaining equipment and services. Some companies will bill a fixed customer charge, such as $1.00 per meter, each month.

Special meter charges may be included to pass on to the customer the cost of any special meters, such as time-of-day meters, usually at a flat rate per meter.

Taxes assessed by federal, state, county or city agencies may be added. These are usually a percentage of the total of all other charges.

A minimum bill may also apply.

In some rate structures, individual charges may be figured separately. In others, a few charges may be combined, as when the rate used to calculate the consumption charge may be determined by the level of monthly demand. For instance, if the electricity used is equivalent to or less than 50 hours' use of maximum demand, it may be billed at $0.015/KWH. If it is greater than 200 hours' use of maximum demand, it may be billed at $0.013/KWH. In this way, demand and consumption charges are combined. **See figures 3-27 — Sample electricity bill and SP-12 — Power and light schedule PL-1.**

Applying the right rate structure to the required quantities of fuel or electricity figured in the energy performance estimate is straightforward but depends on the complexity of the rate structure used. The various charges must be calculated and totaled for each billing period. These results are added to produce the annual total. Each charge may have its own type of rate. For instance, the consumption charge may be computed according to a more complex rate type, while fuel adjustment charges may be at a fixed rate per unit of fuel or energy used. Since every company has its own rate structure, it is best to discuss it ahead of time. The building's supplier is the best source.

Straightline rates are simplest as only one price is charged per unit. In this case, the calculation is simply:

cost = price per unit × no. of units

If the cost/unit is 42¢/gallon, the cost of 100 gallons would be:

cost = 42¢/gallon × 100 gallons = $42.00

Stepped rates are more complex. The price charged per unit is constant, as in straight line rates, but depends upon the price "step" within which the total usage falls. The calculation is still:

cost = price per unit × no. of units

Example:		
	0- 50 gallons	46¢/gallon
	51-100 gallons	42¢/gallon
	101-200 gallons	40¢/gallon

the cost of 100 gallons is:
cost = 42¢/gallon × 100 gallons = $42.00
while the cost of 101 gallons is:
cost = 40¢/gallon × 101 gallons = $40.40

(Stepped rates are illegal in some states.)

Block rates are the most complicated. The price charged per unit changes with each succeeding "block" into which consumption falls. The cost for usage falling within each block must be calculated separately and all block costs totaled to arrive at the actual charge. It is a common type of rate structure. The calculations are the same as above, but there are more of them.

cost = price per unit first block × no. of units (first block)
+ price per unit second block × no. of units (second block)
+ ...

Example:	Block	Price
first	50 gallons or less	46¢/gallon
next	50 gallons	42¢/gallon
next	100 gallons	40¢/gallon

The cost of 100 gallons is then:
cost = 50 gallons × 46¢/gallon = $23.00
+ 50 gallons × 42¢/gallon = 21.00
$44.00

The cost of 101 gallons is:
$44.00 + (1 gallon × 40¢/gallon) = $44.40

Reducing unit price with each step or block seems to encourage greater energy consumption. Consequently, some companies are setting up rate structures in which price per unit goes up with usage.

An example of a complete **energy cost calculation for electricity** is included in the sample problem **(figures SP-49, 50).**

Results are easily verified, and this should be done. If the energy planner is satisfied that consumption and demand figures from the energy estimate are accurate, the utility or fuel supplier usually agree to check energy cost calculations. Results can also be compared to past bills. It is important to have confidence in the energy costs calculations, since the energy planner will use them to estimate possible savings when evaluating opportunities for improving energy performance.

If the energy estimate's figures are expressed in BTU's or BTU/hr, they must be converted to the equivalent amount of fuel required, using the applicable conversion factor.

Sample Electricity Bill

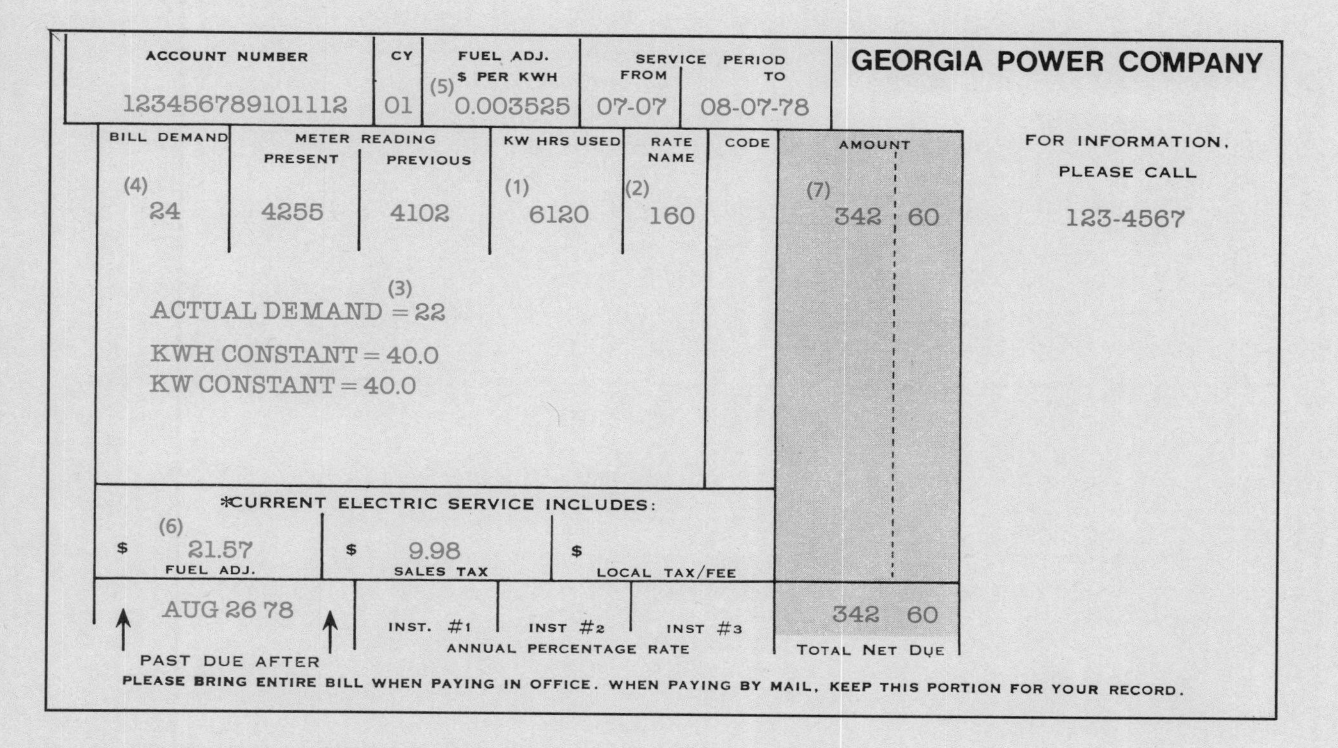

ACCOUNT NUMBER	CY	FUEL ADJ. $ PER KWH	SERVICE PERIOD FROM	TO
123456789101112	01	(5) 0.003525	07-07	08-07-78

GEORGIA POWER COMPANY

BILL DEMAND	METER READING PRESENT	PREVIOUS	KW HRS USED	RATE NAME	CODE	AMOUNT
(4) 24	4255	4102	(1) 6120	(2) 160		(7) 342 60

FOR INFORMATION, PLEASE CALL 123-4567

(3) ACTUAL DEMAND = 22

KWH CONSTANT = 40.0

KW CONSTANT = 40.0

*CURRENT ELECTRIC SERVICE INCLUDES:

$ (6) 21.57	$ 9.98	$
FUEL ADJ.	SALES TAX	LOCAL TAX/FEE

AUG 26 78 — PAST DUE AFTER

INST. #1 — INST #2 — INST #3 — ANNUAL PERCENTAGE RATE

342 60 — TOTAL NET DUE

PLEASE BRING ENTIRE BILL WHEN PAYING IN OFFICE. WHEN PAYING BY MAIL, KEEP THIS PORTION FOR YOUR RECORD.

(1) Consumption in KWH (the difference between the present and previous meter readings multiplied by the meter constant).
(2) The rate classification designating the rate on which a customer is billed.
(3) Highest actual demand registered on the meter during the current service period in KW.
(4) The demand used to bill the account in KW. This may be different from (8) if the electric rate schedule includes a "ratchet" clause.
(5) Fuel adjustment rate in $/KWH used.
(6) Fuel adjustment charge [(5) × (1)].
(7) Total of consumption, billing demand and fuel adjustment charges.

Figure 3-27

Identifying Opportunities for Improved Energy Performance

Practical opportunities for improving a building's performance may be identified in a two-part sequence. Both parts involve the prior audit and/or analysis of the building's *present* performance. The first part consists of reviewing opportunities already identified during the audit or analysis. The second part involves a methodical look for added opportunities using the audit and/or analysis as the basis for search.

The owner, operations and maintenance staff, users, as well as energy suppliers, product manufacturers and contractors may have offered suggestions during the audit and analysis. Numerous publications with energy efficiency checklists and ideas are available as sources of ideas (see References).

The most useful tool is a review of data and calculations performed in the energy analysis. This analysis will not only show monthly energy consumption and peak demand totals, but also the effect of the various loads, operation

Introduction

The key to identifying opportunities is to uncover those with a rewarding return. The energy planner should first record energy reducing opportunities already identified from discussions with owners, previous experience, energy audit and analysis, product manufacturers and from publications and checklists. Then the planners should look for added opportunities by thoroughly reviewing the data contained in the energy audit and/or analysis.

Listing opportunities already identified

Ideas may occur throughout the process, starting with initial conversations with building owners and their operating staff. It is good practice to **record all ideas** to maintain good relations with all parties, even though some of the ideas may not eventually be operable.

Owners sometimes acquire fixed notions as to solution of building energy problems. Thus, energy planners repeatedly encounter certain ideas. These may include:

- Solar energy
- More insulation
- Double-glazing
- Computer control systems
- Weatherstripping
- Changing from incandescent to fluorescent lights
- Natural ventilation with operable windows

In some buildings, these individual items may not be feasible. Some may actually increase energy consumption. On occasion, all may be feasible. **The key is to use the solution appropriate to the specific building's problems.**

Building users generate ideas that mainly have to do with comfort. Comfort is, after all, the reason for the building's existence, so they should be listened to. Their comments will point to areas of the building where systems may not be performing properly. If an area is too warm while another is too cold, balancing the air distribution systems may save energy as well as improve comfort. If users complain about glare, reduced lighting levels and relocation or changing of fixtures should be explored.

The energy audit and energy analysis as sources of opportunities

During the energy audit and/or analysis, one can ask, "what if..." questions about potential modifications. If solar heat through glass is suspected as a culprit, one might ask, "what if the shading coefficient were reduced to 0.35 from its current 0.70?" Such questions about the characteristics of materials or equipment stimulate the search for energy-saving improvements, along with means of carrying them out.

For example, if solar heat gain through windows or excessive infiltration was noted during the energy analysis, the energy planner may describe an ideal set of performance characteristics for a modified window design and look for methods and materials to modify the existing windows or replace them. These methods can then be placed on the list for evaluation.

Product manufacturers as sources

Since higher energy costs and resulting concerns have created expanded markets for manufacturers, there are now many new products to improve a building's energy performance. These products range from supplemental glazing systems added to single-glazed windows to produce double-glazing, to improved solid state controls for multi-zone HVAC systems.

The energy planner should **first come up with ideas from the energy audit and analysis, and second with ideas from manufacturers. That is because the analysis will provide a framework against which to evaluate the manufacturers' products.** In the shading coefficient example, there are fewer products which meet a clear set of specific performance criteria than there are devices which can be added to windows. Establishing performance criteria will help the energy planner quickly to sort out those products most worthy of investigation, as well as provide "performance specifications" for bidding purposes.

hours and equipment performance characteristics that created them. With this information, the energy planner can see what contributes most to the building's peak energy demand and average energy consumption, and uncover what can be done to improve performance.

Proposed improvements are then listed according to how well they meet the owner's criteria. This summary of opportunities is useful to prepare for the evaluation step which follows.

Publications and checklists as sources

Publications and checklists are useful in seeing to it that nothing has been forgotten. Still, **most ideas will not be appropriate for a particular building, and should be reviewed in that context.**

Professional publications carry useful articles about individual experiences with particular energy modifications. Again, they may not apply to your building. Keep track of time and budget when reviewing and cataloguing publications, as it can quickly become a full-time task. It is best to rely on analysis to identify the main areas of concern, and then merely look to publications for examples.

Looking for additional opportunities

The energy planner should review calculated loads from the audit and/or analysis, as well as the relationships between the loads and with the system response. Final results of energy performance calculations of themselves will not help in the absence of a broad stroke understanding of the building's energy performance. It is essential that the energy planner note the key variables that determined the building's energy use during the energy analysis.

While chances for improvement may come from any source, the biggest ones will emerge by methodically **reviewing the audit and/or analysis.** This is done by means of a series of questions. These vary from one project to another. They are best illustrated by referring to a specific building, in this case the sample building.

First arrange the owner's concerns in order of priorities. Then uncover factors that influence these concerns in order of their importance.

Identifying the owner's primary concerns

Question: What is the owner's primary concern?

Answer: Reducing energy costs in this all electric building.

Question: What determines the building's electric costs?

Answer: Using the building from the sample problem and looking at **figure SP-51** gives the following estimates:

45% of energy costs are due to energy consumption. In the sample problem's rate schedule, the owner can save between 1.3¢ and 1.5¢ on the first 462,546 KWH savings in January, for instance, and 3.8¢ for the next 300,000 KWH savings, due to the effect of demand on the rate structure. Unless the demand is changed the **first savings are at the "cheap" end of the scale. The more the owner saves, the more substantial the savings become.** Conversely, the first savings with this rate schedule are only a fraction of the average unit cost of energy. This seriously discourages energy saving.

6% of costs are due to fuel adjustment. This amounts to 0.227¢/KWH and is directly related to energy consumption, regardless of how much is used.

45% of energy costs are due to peak demand. The owner's yearly peaks are in summer and fall. Roughly $50/kilowatt will be saved for a reduction in the peak.

4% of energy costs are due to taxes. These are added to the total costs.

The energy planner can see at this point that reducing consumption one kilowatt hour saves 1.3¢ to 3.8¢, whereas saving a kilowatt of demand at the time of the building's peak saves $50.

Consumption and related fuel adjustment charges make up 51% of the total for the sample problem building. Therefore, investigate next those factors that contribute most to the building's energy consumption.

Identifying the largest sources of energy consumption

Question: To which factors is the building's energy consumption sensitive?

Answer: The answer is found by breaking down energy consumption of the building's major users of purchased energy and listing variables which affect energy consumption of equipment involved. The sample problem discloses eight energy users, each with a set of variables **(see figure SP-48).**

These quantities are based on a particular building design in a particular climate. If window area or lighting levels change, for example, these quantities change substantially.

After identifying these causes, the energy planner investigates further the largest causes of energy consumption.

Question: What modifications can be made to reduce consumption of the most significant item?

Answer: Since artificial lighting accounts for 31%, further identify major variables that effect its consumption. Thus:

"Hours of operation" and "occupancy"
Can total hours of building operation be reduced? Or can they be reduced in any unoccupied areas, such as storage rooms, equipment rooms or outdoor areas? Can the building operator turn lights off and on more quickly? Can cleaners do their work in daytime hours—or faster?

"Tasks involved"
Are lighting quantity, quality and design appropriate for the task?

"Fixture placement"
Is there glare? Can the number of lighting fixtures be reduced by more efficient placement?

"Efficiency of system"
Can more efficient lamps, ballasts and transformers be used?

"Availability of daylight"
Can any room be adequately lit with daylight? If so, for what periods?

The other boosters of energy costs can then be identified in the same way.

% of total consumption due to:	***Sensitive to:***
Artificial lighting (31%)	*Hours of operation/occupancy* *Tasks involved* *Lighting fixture placement* *Efficiency of system* *Availability of daylight*
Heating (30%)	*Hours of operation* *Outside air that must be heated in cold weather required for ventilation by building code* *Outside air entering due to infiltration through wall, doors and openings* *Transmission losses* *Room temperatures maintained* *Distribution losses* *Solar heat gain* *Internal sensible heat gains (lights, equipment, people)* *Area of space requiring heating* *Efficiency of system*
Fan coil fans at perimeter (15%)	*Hours of operation* *Efficiency of system* *Occupants' setting of fan speeds*
Air-conditioning compressor (8%)	*Hours of operation* *Sensible heat gain from lighting, sun, equipment, transmission through walls* *Latent and sensible heat gain from people and outside air as required for ventilation by code* *Maintained room temperatures and humidity* *Infiltration* *Efficiency of system and set points at controls* *Space cooling loads*
Miscellaneous HVAC fans and pumps (7%)	*Hours of operation* *Distribution system losses* *Efficiency of system*
VAV fan (4%)	*Ventilation requirements* *Hours of operation* *Distribution* *Efficiency of system*
Hot water (2%)	*Water temperature supplied for use (hand washing, primarily)* *City water temperature* *Distribution method* *Efficiency of system*
Elevators (2%)	*Frequency of use* *Hours of operation* *Efficiency of system; program of operation and controls for motors*
Other (1%)	

Identifying the largest sources of energy demand

Question: What determines peak demand charge?

Answer: Demand in the sample problem building is greatest in January, but because the utility has its greatest system-wide peak in the summer, billing demand is actually determined by 95% of peak summer demand, which occurs in September, or else 60% of peak winter demand **(figure SP-53).** In the sample problem, summer demand is most significant so it must be investigated first. Then, if it can be reduced, examine winter demand further.

Question: What are the major contributors to the summer peak demand and what variables offset their peak energy use? **(figure SP-52).**

Answer: In the sample problem building, summer peak demand is in the afternoon on the hottest clear day, most likely on Monday or after a holiday when attendance is high and all equipment and lights are on. That is also when building systems must work more to cool down the building which has been allowed to "float" over the weekend. Occupants returning to the building from lunch are using elevators at their maximum rate.

Question: What modifications can be made to reduce peak summer demand? (Ideas for reducing energy consumption may also apply to reducing peak demand. These ideas are especially rewarding since they will bring down both components of the electric bill. After identifying modifications, consider how easy it is to make the modification and how significant are the results.)

Answer: 47% of peak summer electric demand is due to HVAC equipment operating to meet maximum thermal load on a hot clear day. The thermal load includes:

"Heat gain through transmission"

Is this amount significant? Probably not for this building due to enormous internal heat gain from lighting. Can this be reduced easily? Are savings worth the effort?

% total summer peak demand due to:	*Contributing factors:*
HVAC (47%)	*Outside ventilation air required by codes is high in temperature and humidity* *Heat gain through transmission* *Maximum solar heat gain* *Maximum internal gains (lights, equipment, people)* *Normal room temperature and humidity* *Accumulated heat, possibly from shut down over weekend* *Efficiency of system is low*
Lights (40%)	*Greatest expected number of lighting fixtures switched on*
Elevators (7%)	*Maximum use of elevators as people leave and enter building* *Efficiency of system*
Hot water (4%)	*Supply water temperatures maintained as usual; water heater may have come on during peak*
Other (2%)	*Normal operations* *Efficiency of system*

"Maximum solar heat gain"

Can solar heat gain come down through use of shading devices, reflective glass or coatings? Which glass orientation receives the most severe impact? Are there other benefits to reducing heat and light through glass?

"Internal heat gain"

Can some lighting fixtures or office equipment be turned off during or just prior to peak periods to reduce electrical draw as well as space cooling loads?

"Normal room temperature and humidity maintained"

Are the present conditions the most comfortable? Can less comfortable conditions be tolerated for short periods without unacceptable discomfort?

"Capacity of system"

Can equipment such as chillers and cooling towers be modified or cycled off and on for short periods?

"Efficiency of system"

Can the HVAC system's efficiency be improved?

40% of peak summer demand is due to lighting fixtures being switched on as usual.

Can any lighting fixtures be turned off during peak periods?

7% is due to maximum operation of elevators when all users are entering or leaving the building.

Can some users be encouraged to use stairs? Can work schedules be staggered so everyone does not need to leave or enter the building at the same time? Can elevators be reprogrammed to reduce demand if above is done?

4% is due to normal water heating.

Can the water heater be turned off during peak periods? Are there more efficient water heaters? Can another source of energy be used to heat water?

Question: What are the major contributors to winter peak demand and what causes their peak energy use?
(figure SP-52)

Answer: In the sample problem building, peak winter demand is on the coldest winter morning, before heat gain from people, lighting fixtures and office equipment begins to be felt, but after the VAV system has begun to deliver 60°F ventilation air to exterior zones. Elevators are operating at maximum as people come to work and lighting fixtures and office equipment have just been turned on. A review of winter recording demand charts would confirm the time of peak winter demand.

% total peak winter demand due to:	**Contributing factors:**
HVAC (63%)	*Coldest outside air temperatures* *No solar heat gain (dark outside since peak is probably a cold, winter morning)* *No internal heat gain (lights, equipment and people have not begun to contribute their heat)* *Heating of cold outside air being used for ventilation as usual* *Heat loss through transmission* *Heat loss through infiltration* *Normal room temperature maintained* *Capacity of system* *Efficiency of system*
Lights (28%)	*Greatest expected number of lighting fixtures is on*
Elevators (5%)	*Maximum use of elevators as people come to work* *Efficiency of system*
Hot water (3%)	*Supply water temperatures maintained* *Heater on as usual* *Efficiency of system*
Other (1%)	

Question: What modifications can be made to reduce peak winter demand?

Answer: 63% of winter peak electric demand is due to HVAC equipment operating to meet maximum thermal loads on the coldest, darkest day of the year. In the early morning, the building does not benefit from heat gain from the lighting fixtures, equipment and people.

Consider:

"Heat loss through transmission"

Can this be reduced?

"Heat loss through infiltration"

Can this be reduced?

"Heating of outside air being used for ventilation as usual"

Can ventilation and exhaust be shut off during peak periods, especially in the morning when the building is heating up but not yet occupied?

"Normal room temperatures maintained"

Can cooler room temperatures be tolerated during peak periods? Could the building's users wear added clothing or will they bring in energy consuming space heaters?

"Capacity of system"

Can the boiler be shut down entirely during peak periods?

"Efficiency of system"

Can a more efficient boiler be used, or boiler efficiency increased? Can a different energy source be used for heating?

28% of peak winter demand is due to lighting fixtures switched on as usual.

Methods for reducing lighting are the same as when reducing summer peak demand.

5% of peak demand is due to maximum operation of elevators as people come to work. Refer to summer peak demand discussion.

3% is due to normal water heating. Refer to summer peak demand discussion.

Examining all significant variables

Another useful approach to help identify opportunities once major sources of energy costs are identified is to consider four ways in which costs can be reduced:

Reduce the job to be done (the initial need for the energy).
Reduce the time spent doing the job (how often that need is met).
Increase efficiency with which the job is done.
Reduce the cost per unit of energy used to do the job.

The formula is:

Energy costs =

$$\frac{\text{Job to be done} \times \text{Time spent doing job} \times \text{Cost per unit of energy used}}{\text{Efficiency of energy use}}$$

Changing the elements in this equation offers opportunities for energy cost savings, for example:

Reducing the job to be done

Energy consumption for the building's equipment varies with the loads. If these can be reduced, the amount of energy needed to operate the equipment can also be reduced. The equipment may even be eliminated in the first place. Consider, then, the following:

If proposed construction, does the building need to be as big as it is?
Can functions be combined, schedules rearranged or areas closed off?
Are areas requiring similar climate control grouped together?
Does the building have a perimeter/interior ratio appropriate for its predominant loads?
Can unconditioned areas buffer conditioned areas to lessen the impact of climate?
Can sunlight and natural air movement be exploited through rearrangement of interior spaces?
Can work-tasks that require the highest lighting levels be relocated to areas with good daylighting?
What is the potential of natural versus mechanical ventilation?
What is the potential for waste heat recovery?
Is active solar heating applicable?

In many building types artificial lighting and HVAC equipment are usually the the best candidates for energy savings. Lighting and HVAC loads are determined by weather; by building area, mass, envelope configuration and construction; by building use patterns; and by numbers of people, and activities. This is discussed in the "Introductory Concepts" Chapter 1. A few examples illustrate the point.

Heat transmission load depends on envelope construction, surface area, U-factors, mass and reflectance. It may be possible to reduce exposed surface areas, add insulation, or reduce the difference between interior and outside temperatures in cold climates.

Infiltration load is caused by air leaks through the building envelope and may depend on wind velocity and sometimes stack effect, as well as differences between inside and outside temperatures and air pressures. It may be possible to plug or reduce the size of leaks with caulking, weatherstripping, new gaskets or new windows.

The load on equipment due to **solar heat gain** depends on the amount and duration of solar heat striking the building's surface, especially on glazed window openings and skylights. Load varies with sun intensity, altitude and bearing, cloud cover and haze.

Direct solar gain may be an asset or a liability, depending on how the building has been designed to receive it. It also depends on the amount of heat transmitted through glass, as indicated by the glass shading coefficients. Heat build-up in opaque materials is influenced by surface reflectivity. Solutions could include shading the building's surfaces to intercept the sun, reflect the sun, filter out the sun's heat, or change the surface area exposed to the sun. It may be possible to do this so as to permit solar heat gain when it is beneficial and restrict it when it is not.

The load on equipment due to heat gain from **lighting fixtures** and equipment is a function of the heat they give off, the number of lights or pieces of equipment and how much of the time they are used. One could decrease heat gain per lighting fixture or piece of equipment by replacing it with a more efficient unit, by exhausting some of the excess heat or by reducing the number of lights and pieces of equipment. Perhaps they could be operated less often, or there may be an effective way to make use of this heat.

The load due to **latent heat gain** depends on the amount of moisture to be removed from the air to reach the indoor standard, and on the amount of air to be conditioned. The energy planner should see if the primary source of latent heat gain can be reduced on humid days, if uses that require strict humidity control can be grouped together, and if internal sources of moisture can be hooded and the moisture exhausted, as in the case of kitchens and laboratories.

Reducing **ventilation supply air** can reduce the amount of air that must be circulated, as well as accompanying distribution and friction losses leading to a lower consumption of fan energy. It may be possible to group activities which require extra odor, fume or smoke control so they are separate from spaces requiring little ventilation, such as storage and unoccupied spaces.

The consumption of **non-HVAC equipment** may also vary with loads. For instance, the load on the domestic hot water heater depends on the amount of water to be heated, storage available and required temperature rise. One should explore to see if the amount of water to be heated may be reduced by cutting back on hot water flow, by reducing distribution losses with insulation, or by restricting storage losses with added insulation. Perhaps the required temperature rise and storage losses could be reduced by lowering output temperature.

Reducing time spent doing the job

If equipment is less often used, it may not use up as much energy, except if the equipment is cycled in response to loads.

Examine use and operation schedules during the energy audit or analysis. Many building operators, for example, have begun to shut down air conditioning sooner before they leave, figuring it takes a little while for conditions to become uncomfortable.

Turning items on and off may save energy by reducing total operation hours, but some equipment is designed to run best when in continuous operation.

Equipment may be more efficient at partial load than at full load. Turning it on and off, thereby forcing it to operate at full load and then stop, may raise energy consumption instead of reducing it. If not designed for frequent starts and stops, it may wear out faster. Equipment manufacturers should be consulted as to the best operating method.

Other HVAC-related items to consider include:

- Reducing the hours a building is air-conditioned and/or ventilated.
- Reducing the hours that exhaust and ventilation fans operate.
- Using nocturnal cooling, passive solar, and waste heat to reduce equipment operation hours.
- Cycling certain pieces of equipment instead of running them continuously, especially if not needed all the time.
- Reducing the amount of time required for artificial lighting.

Another method of reducing equipment operation is to explore **changing habits** of those who use and operate the building. If this is tried, the energy planner may need to become both an educator and a politician.

Best **operation and control** of a building and its equipment means using the least amount of energy to do a job while conforming to performance standards. A thorough energy analysis includes recommendations for an efficient operations and control plan.

Experience shows that a good **operating engineer and staff** can be the greatest energy-saving asset a building has. Some building operators always look out for local weather conditions and their own experience with the building in order to decide when to start up systems. This may be the most efficient approach. If a good operator is not available, automatic operation and control systems may be useful, however.

Control systems on the market range from large scale computer systems which control many aspects of a building's operation to simple time clocks and visual observation by operations staff. When considering control systems, note:

- The building may not need a large sophisticated system.
- Programming or wiring the system may cost more than the hardware.
- The energy analysis should be checked to see if the manufacturer's claims are credible.
- Automatic control devices require regular recalibration.
- The more complex the device, the more likely are malfunctions, high maintenance and repair costs.
- The impact of operation and control strategy on life expectancy of the equipment should be checked.

Whatever the method, operation and control procedures should be reviewed regularly to make sure they are still effective.

Increasing efficiency with which the job is done

The energy planner should look into the relationship between loads, system response, flows of energy (including waste heat) and equipment and energy consumption, using the energy audit and/or analysis as a base.

One may do several things to improve efficiency of a piece of equipment, including replacing it if worn down. Questions to ask include: Is the equipment kept in good repair and operating close to ideal condition? When considering replacing worn out equipment, is there a more efficient unit, because, in reponse to new stress on energy conservation, manufacturers are developing more efficient equipment? Can sloppy out-of-date controls be modified for greater precision? If there is a power factor charge, can the equipment's power factor be improved by installing capacitors?

Reducing cost per unit of energy used to do the job

There are two ways to reduce cost of energy per unit used: find a less expensive energy source, and/or use energy in a way to **take advantage of the rate structure.**

If a demand charge or time-of-day billing is incorporated in the utility rate structure, big savings are possible by changing patterns of energy use.

Reduced peak energy use may be accomplished simply by starting the building's equipment in steps rather than all at once. Non-essential energy uses can be done away with during peak hours. Users can alter their times of arrival and departure if this seems in order. Heat or coolness can be generated during off peak hours and stored for later use. Ways to anticipate needs before peak demand periods occur should be explored, and equipment run accordingly.

Various demand management systems and load shedding devices are being sold for turning off loads not absolutely needed during peak periods. These systems may be very complex. They monitor energy consumption to predict peak periods and turn off equipment according to a prearranged program.

If the utility company uses a time-of-day rate structure, consider taking care of as many of the building's needs as possible during lowest rate periods. This may include changing the hours of building use.

Ways to manipulate energy use to take advantage of rate structures depend on the local conditions under which energy is supplied. The energy planner should study these with great care.

Evaluating Opportunities

The opportunities identified in the previous chapter must be evaluated so the most rewarding ones may be chosen and carried out. The energy planner and owner must decide which aspects of the building's performance to consider and what selection criteria to use. The energy planner will then size up the impact of carrying out each opportunity, estimate the costs, compare results with the selection criteria, and recommend for or against it.

Typically, the process calls for more than one cycle of evaluation. The first evaluation may not produce enough information about some opportunities. Also, selection criteria may be changed after results of the first cycle are reviewed.

Two typical cycles of evaluation are discussed: a "quick" evaluation and a "detailed" evaluation. In a "quick" evaluation, each opportunity's impact is sized up by a quick look at the energy analysis to see how those results would change. A rough cost estimate is made, and results compared with the selection criteria.

Introduction

It is good to approach the evaluation of opportunities systematically. Although strict adherence to the methods described below may strain the budget, the underlying approach at least should be used.

Typically, the evaluation process goes through more than one cycle, for two reasons.

Energy planners begin to size up potential opportunities from the moment they walk through a building. However, even though "quick and dirty" evaluations will indicate at once that some opportunities are feasible, they may not point in a clear direction for others. Evaluations then involve greater and greater levels of detail.

The first evaluation criteria discussed with the owner are not likely to be the last. That is the second reason why the evaluation is cyclical. The criteria developed in the contract discussions should be looked at again as this phase begins, as the owner may now bring in new criteria. For example, even though laying out fluorescent lighting to match the task below may meet the owner's earlier criteria, the energy planner may now learn the owner likes the present ceiling grid because it "looks organized," and the lighting must remain as is.

Evaluation procedure

*The **flowchart (figure 5-1)** reflects the basic evaluation procedure. It shows that any number of cycles may be required before a recommendation is made. Usually, two cycles are enough. At times, feasibility may come into question even after a detailed evaluation. In that case a more thorough analysis is required. This should be brought out. The owner should always realize there may be a margin of error behind each prediction and recommendation.*

The effort spent on this phase is set by the project's budget and schedule and will determine the number and detail of evaluating cycles. In some cases, the owner may want to contract up through a quick evaluation only, review progress, and enter into another contract for detailed evaluation. A quick evaluation will not call for much added investment by the owner beyond the prior energy analysis and opportunity finding. A detailed evaluation, however, can become complex and the owner may want to review the entire project before committing such extra resources.

During evaluation, the best tool is, as before, the energy analysis. Evaluating a modification amounts to changing the analysis to reflect its impact. If the energy analysis has not been thorough enough to permit this, it may need to be redone in greater detail:

Final presentation to the owner may call for a written report. The energy planner may find clear descriptions of this type of work hard to write, so adequate time should be allowed.

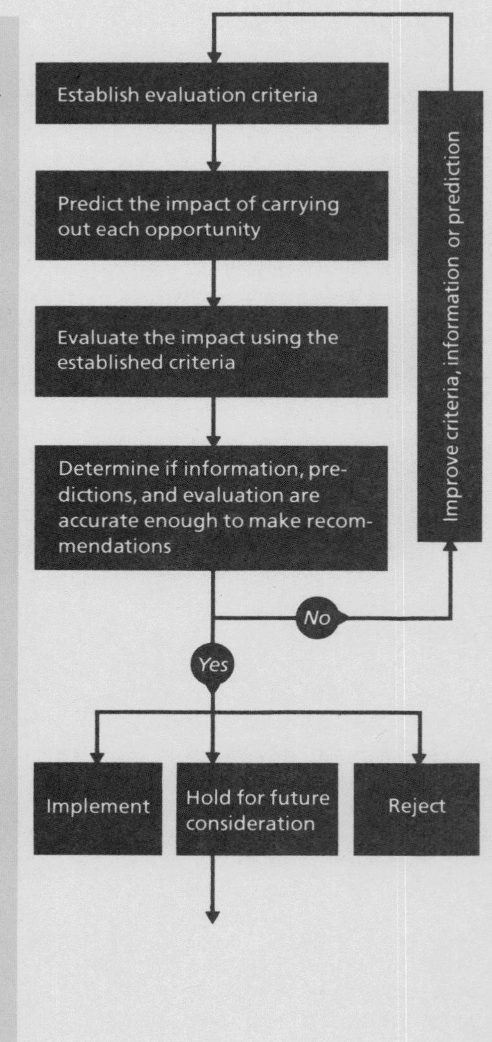

Figure 5-1

Some opportunities will clearly be worth carrying out after only a quick evaluation; others will call for a more detailed investigation. All recommendations should be thoroughly reviewed by the owner and any changes in evaluation criteria firmed up before beginning a "detailed" evaluation.

In the "detailed" evaluation, the performance estimate prepared for the energy analysis is actually redone using new input data to reflect changes caused by the modifications being considered. This is required to predict accurately the impact of many proposed modifications, especially those which affect more than one factor influencing the building's energy use. The main difference between a "detailed" and a "quick" evaluation is this accurate figuring of the results of changes by reworking the energy analysis estimate.

It is hard to draw firm lines between the two methods but, in general, cost-benefit studies, comfort predictions and the impact of a modification in every area are done more accurately for the "detailed" than the "quick" evaluation.

Performing a quick evaluation

The quick evaluation uncovers those opportunities clearly worth doing, those to be rejected and those in need of more detailed evaluation. It also clarifies what to do to evaluate a modification more accurately. Research and calculations should be kept to a minimum. A review of the energy analysis should be enough, along with rough cost estimates and a short look at each of the other evaluation criteria. If more accurate information is required, the modification should be set aside for detailed evaluation and the owner's approval obtained before spending any more time and effort on it.

Establishing evaluation criteria

Before evaluating an opportunity, the energy planner must define the criteria an opportunity must meet. Criteria should have been discussed with the owner ahead of the energy audit and analysis, and should now be reviewed once more in light of results. Key considerations are listed below. **The list changes from project to project** and is by no means complete, but provides for initial guidelines.

Savings

Energy

What is the project's energy performance goal? How much will each opportunity improve the project's energy performance, of itself and when combined with other opportunities?

Dollar savings

How much does the owner wish to reduce annual operating expenses? How much will each opportunity save of itself and when combined with other opportunities?

Costs

Initial costs

How much is the owner willing to pay initially to meet goals?

What are the initial costs of carrying out the opportunity, including design costs and costs such as construction, administration and financing?

Continuing costs

What will be the cost with respect to maintenance and replacement of equipment and materials which are modified? Under-used equipment deteriorates more slowly and needs less maintenance and replacement.

What will be the effect of inflation on projected costs stemming from the modifications? Interest costs are inflation free; energy costs inflate rapidly.

What will be the effect on taxes? Will the energy efficient modifications be taxed as operating expenses or as capital investment? How fast can the investment be depreciated? Is there a tax incentive?

Costs vs. savings

What costs vs. savings ratio is acceptable to the owner? If the owner wants a very short "payback" period, the number of items to be investigated and the scope of the analysis are reduced.

How will this ratio be determined? For example, by simple payback period of x years, discounted payback period of x years, or with a life-cycle cost study?

Comfort

What will be the effect of comfort? Will occupant complaints increase if operating temperatures are changed? Reducing hours of operation can reduce comfort during off-hours. Is that acceptable? Some owners may have as a goal the improvement of comfort.

Functional impact

What will be each modification's impact on the way the building fulfills its purpose? Will functional patterns have to change?

Environmental impact

Will the modifications reduce emissions of pollutants and thereby better meet regulations or standards?

Visual impact

Does the "image" of the building improve? Are employees' surroundings made more attractive? Energy-efficient redesign of a building may become a part of a facelift or renovation of an existing building.

Code and regulation acceptance

Must the building be brought into conformance with any regulation or code?

Predicting the impact of each modification

The best approach is to ask **"What is the greatest benefit I could hope to gain?"** and use the available data to estimate the savings as easily as possible.

*For example, assume that you have completed the detailed energy estimate for the sample problem at the end of this book, so this data is available, and that a proposed modification is to install shading devices that would **cut the solar heat gain through windows by 50%** during both summer and winter. There are two obvious results of such a reduction in solar heat gain:*

- *A The **refrigeration load is reduced** when there is a cooling load.*
- *B The **heating load is increased** if the outside temperature is cold and there is a heating load.*

There are also several not-so-obvious possibilities:

- *C The reduction in peak cooling loads might permit the HVAC system perimeter **fan-coil units to be operated on medium fan speed instead of high speed.***
- *D If the shading device greatly reduces the visible light transmission, it may also **reduce glare and permit a reduction in lighting power** input. In many buildings sunlight entering through windows produces so much glare and such a high brightness ratio that lighting levels must be increased to compensate.*

Item A–The **direct reduction in refrigeration load during the cooling season** affects both peak demand and energy consumption. The peak demand estimate in **figure SP-17, Peak central system cooling load, and SP-24, Peak cooling demand profile** can be used to estimate the peak demand reduction.

The minimum demand charge for this building is being set by the peak demand during the summer months, June through September. **Figure SP-28, Monthly peak demand summary,** shows a maximum value of **1184 KW** in September **which occurred during the heating condition** on a cool morning. **Unless this heating peak is reduced by locking out most of the boiler capacity and preventing it from operating until October 1, changing the shading coefficient of the glass will have no effect on the building demand charge.**

If cooling then sets the minimum demand; it can be reduced with the solar device. The glass solar load of all four exposures shown in **figure SP-17, Peak central system cooling load,** totals 1,505,934 BTUH, or 14.6% of the 10,328,774 BTUH total refrigeration load. Cutting the solar gain by 50% would reduce the total refrigeration load by 50% x 14.6% or 7.3%. **Figure SP-24, Peak cooling demand profile,** shows the refrigeration compressor peak demand power input as 641 KW, which would thereby be reduced by 7.3% x 641 or 47 KW.

The average energy savings during the cooling season is established by estimating the savings at an average condition and multiplying by the hours of operation. The majority of the refrigeration system energy consumption occurs when the outside temperature is 50° or higher. From **figure SP-66, Average annual HVAC consumption,** there are about 3000 hours during the occupied period in the 52° through 92° temperature bins. **From figure SP-35, Occupied period average energy profile,** the refrigeration compressor power at 72° is approximately halfway between the 52° and 92° points, so 72° is an average cooling season operating condition. **Figure SP-31, Average block load,** shows the 72° average load, from which the glass solar load is 735,912 BTU, or 20% of the 3,685,946 BTUH total cooling load. Reducing the solar gain by 50% would reduce the load by half of 20%, or 10%. **From figure SP-35,** assuming that the load on both the compressor (218 KW) and cooling tower fans (54 KW) would be reduced, the total reduction is 10% x 272 KW, or 27 KW. The annual savings are 27 KW x 3,000 hours = 81,000 KWH.

Item B–The increased boiler consumption that occurs because of the reduced beneficial solar heat gain through windows during cold weather can be quickly estimated by reference to **figure SP-66, Average annual HVAC electric consumption.** The annual credit for solar gain on weekends is shown as 264,712 KWH. **If the glass shading coefficient were reduced 50%, this credit would be also reduced 50%, thereby increasing the boiler load by 132,356 KWH.**

Estimating the effect during occupied hours would require recalculation of the many hour-by-hour, zone-by-zone estimates such as that shown in **figure SP-32.** The complexity of this revision would require postponement until the detailed evaluation phase. It should be noted that there will be an increase in energy consumption due to less beneficial solar heat gain, but that some of the increase will be offset by the reduced need for refrigeration during cold weather in those perimeter areas that have a cooling load due to solar gain and internal heat gain.

Item C–The power reduction if the fan-coil units were changed to medium speed operation involves both peak demand and energy consumption. Reference to the fan-coil unit capacity ratings showed that the fan-coil units would probably have adequate capacity at medium speed, and that the power input at medium speed was approximately 17% less than the high speed power input. (The reader should note that the ratio of power input per CFM varies widely depending on motor size, type, speed and voltage, as much as 50%.)

Since all units are operating constantly during occupied hours, the peak demand reduction is easily estimated by multiplying the 153 KW total of all fan-coil units motors by the 17% reduction to obtain a 26 KW reduction in peak demand. The annual energy savings can be approximated by assuming that the operation of the fan-coil units is primarily for cooling when it is above 50° outside, and primarily for heating when it is below 50°. **The reduction in power during heating operation does not save energy because the motor power warms the room and reduces the electric resistance boiler load by the same amount.** The building is occupied approximately 3000 hours when the outside temperature is above 50° **(figure SP-66, Average annual HVAC electric consumption),** so the annual power savings during cooling operation is 26 KW x 3,000 hours = 78,000 KWH.

A side effect of reducing the fan-coil motor power is to reduce the refrigeration load during cooling operation. Figure SP-17, Peak central system cooling load, shows that the heat gain from these motors contributed 522,189 BTUH to the total 10,328,774 BTUH refrigeration load, or 5.1% of the total. Operation at medium speed would reduce the refrigeration machine power input (641 KW at peak demand) by 17% x 5.1%, or about 1%. (See **figure SP-24, Peak cooling demand profile**). This would reduce the peak demand another 6 KW in addition to the 26 KW direct reduction in fan-coil motor input.

The annual savings during cooling operation are estimated by calculating the annual reduction in cooling load and multiplying by the refrigeration system performance. The 26 KW reduction in motor heat x 3413 BTUH per KW divided by 12,000 BTUH per ton = 7.4 tons reduction in cooling load. At an average summer condition of 72° outside temperature, **figure SP-35** shows the compressor power as 218 KW and the cooling tower fan as 54 KW, for a total power input of 272 KW that is variable with refrigeration load. The refrigeration load at 72° was 307 tons **(figure SP-31, Average block load)**, so the refrigeration system performance is 272 KW ÷ 307 ton or 0.89 KW per ton. The annual reduction is 7.4 tons x 0.89 KW per ton x 3,000 hours of cooling season operation, or 19,800 KWH.

Item D – The power reduction due to reduced artificial lighting in the perimeter offices was made possible by the reduction in glare and brightness at the windows. The original lighting design was based on providing enough artificial lighting to overcome this glare and accompanying veiling reflections and shadows, as well as prevent the room interiors from looking gloomy. Suppose that a preliminary check shows that the proposed glass shading device will substantially reduce the glare and permit a 20% reduction in lighting power input without sacrifice in quality of lighting. Like the fan-coil units, lights increase the building's electric use in two ways: through direct consumption and also indirectly because the heat from the lights increases the HVAC refrigeration requirements during the cooling season. When it is cold outside and the rooms need heating, the heat from the lights is useful, and any reduction in lighting power input would be offset by the increased boiler load.

The direct and indirect results of reducing the lighting power could be estimated in the same manner as the fan-coil unit motor power reduction.

An alternative approach is to use the lighting estimates.

Figure SP-43, Energy estimate – lighting, shows that the lighting peak demand is based on a 90% diversity factor for tenant lighting. If the offices are laid out with perimeter and interior separated, then since the installed wattage in the tenant perimeter spaces was 429 KW, a 20% power reduction would save 429 x .9 x .20 x 77 KW.

The sum of the installed wattages in tenant spaces, lobbies, and corridors is 972 KW and their daily consumption is 7774 KWH. The building is occupied 255 days per year, so the annual power reduction of lighting is:

$$\frac{\text{429 KW perimeter offices}}{\text{972 KW total}} \times 7{,}774 \text{ KWH/Day} \times 255 \text{ Days} \times 20\% \text{ reduction} = 175{,}000 \text{ KWH/Yr}$$

The portion of this reduction that occurs during heating operation will be offset, however, so we must correct the reduction by the proportion of cooling hours to total occupied hours:

$$175{,}000 \text{ KWH} \times \frac{3{,}000 \text{ cooling hours}}{4{,}092 \text{ total hours}} = 128{,}000 \text{ KWH}$$

The indirect effect of refrigeration load reduction is calculated in the same manner as the fan-coil motors, for a 17 KW reduction in peak demand and a 32,600 KWH annual savings.

The total impact of reducing the glass shading coefficient and visible light transmission is approximately the sum of the sub-estimates.

		Peak Demand (KW)	Consumption (KWH)
A	Direct cooling load reduction	47	81,000
B	Increased boiler load due to reduced beneficial solar heat gain on weekends	0	–132,400
C	Fan-coil unit speed reduction		
	Direct motor power savings	26	78,000
	Indirect refrigeration savings	6	19,800
D	Lighting intensity reduction		
	Direct lighting power savings	77	128,000
	Indirect refrigeration savings	17	32,600
		173 KW	207,000 KWH

Using the data in **figure SP-51, Average annual electric costs,** the possible demand reduction is about 7% of the summer peak and the energy reduction is 2.5% of the annual total. Since the energy charge is slightly greater (54%) than the demand charge (46%), the possible savings is about (54% of 2.5% plus 46% of 7%) 4.6% of the $300,000 annual cost, or $13,800. If the owner will invest up to 5 times the annual savings, an investment of $69,000 could be justified. Since **figure SP-6, Building data by zone** indicates approximately 65,240 sq. ft. of glass, this budget would allow about $1 per sq. ft. of glass for treatment and lighting reduction, a low amount.

Note that **the increased boiler use for heating more than cancelled the energy savings during the cooling season. Unless items C or D are implemented, there is little economic justification for improving the glass shade factor. It is frequently the case that an indirect saving made possible by a change elsewhere will be greater than the obvious direct saving. This emphasizes the importance of investigating the total impact of a proposed modification upon all of the other building energy consuming systems.**

If the offices in the example were laid out with low partitions in an office landscaping manner, then all occupants with a view of the bright windows could accept a 20% electric lighting reduction. Calculating in the same manner for all the tenant spaces would increase the item D savings from 160,600 KWH to 348,000 KWH. (The savings in the interior zone will not be exactly proportional because while the perimeter offices require heating during a portion of the winter, the interior always requires cooling. However, this cooling is accomplished with an economizer cycle in the winter. The estimate is close enough for a rough analysis.)

The savings would become roughly: demand 11.6% and consumption 4.9% for an approximately (46% of 11.6% plus 54% of 4.9%) 8% total savings, or $24,000. This would justify a simple five-year payback investment of $120,000. This increased savings is due to the layout of the offices as they are affected by the solar glare reducing device. In fact, **it is easily possible to reduce artificial lighting in most office buildings 20% without reducing glare and an additional 20% in spaces with glare reduction. Therefore, a total energy savings of twice this projected amount is usually realistic.**

The feasibility of visible light transmission and solar heat transmission reduction combined with an artificial lighting reduction in the sample building may be summarized as follows:

A 20% lighting reduction with glare reduction is probably not economically worth considering with a perimeter office layout, but may be practical for an open plan office landscaped arrangement. It will be very lucrative if a 40% reduction results as nearly occurred in the AIA Headquarters Building.

Additionally, in new construction, there is a savings **in initial costs** of lighting systems and refrigeration capacity as well as operating costs and implementation which is certainly attractive.

It should also be evident that a glare reduction device that **allowed the winter solar heat gain without overheating the space would further improve the energy use.**

There may be other benefits to applying solar shading devices. For instance, solar shading may do more to improve **comfort than energy use.** Even on cold winter days, solar heat gain can become so intense that occupants find it too hot to work near glass. The building owner may be as interested in correcting this comfort problem as in saving energy. During detailed evaluation the energy planner will want to install an actual demonstration of the proposed device so the owner and users can undergo the impact firsthand.

Shading devices may also **change a building's appearance.** Owners may like this as it tells the public they are taking action, thus improving image. On the other hand, owners may not wish to change the building's external appearance at all, and the energy planner will be limited to shading devices not visible from outside the building. Shading devices attached to heat absorbing glass can also increase glass temperatures and strain the glass to breaking. Again, a test area helps during detailed evaluation accompanied by careful investigation.

Rough cost estimates based on square foot costs, and conversations with suppliers and contractors, should be enough for the quick evaluation.

In addition to these "first costs," consider the effects of such items as inflation, future energy costs, interest, taxes, operations and maintenance.

The impact of interest rates and taxes is usually best estimated by the owner's tax counsel. Each opportunity should be examined to see if it is an operating expense or a capital expense. In a capital expense, counsel should figure how quickly it can be depreciated and whether it qualifies for a Federal Income Tax investment tax credit. Perhaps the owner is better off making all modifications this year for tax purposes rather than extending them in phases over the next several years, or vice versa. Many modifications will be easier to evaluate as they entail no changes in loads.

For instance, the owner may think of keeping the building open seven days a week instead of five. By examining **figure SP-41, Annual energy consumption summary,** total electric consumption due to the HVAC equipment during occupied hours (occupied period + garage fans) may easily be divided by the 4092 total annual occupied hours to yield the fact that average HVAC hourly electric consumption is 750 KWH/Hr. Similarly, during 4668 unoccupied hours when HVAC equipment is operating, it is 518 KWH/Hr. Adding 1632 hours to the occupied period and subtracting the same from the unoccupied period makes for an annual increase in HVAC electric consumption of 479,808 KWH.

By examining **figure SP-43, Energy estimate – lighting,** and multiplying monthly lighting totals for all but perimeter lights by 7/5, an increase in lighting electric consumption of 981,891 KWH per year is realized.

Figure SP-48, Annual total energy summary, shows that extra consumption by elevators, house pumps and office equipment adds 103,774 KWH to the total if weekend use is assumed to be the same as weekday use, the worst condition that could be expected.

Thus, **changing the sample problem building to seven day operation** raises annual electric consumption by 1,464,289 KWH, an 18% increase. Peak electric demand, of course, would not be affected, as it probably occurs on a weekday, nor would it affect the building's comfort or appearance. This modification is thus easier to evaluate quickly.

One other change stemming from increasing the building's hours is that the payback period for any energy-saving modification is shortened. If the base year is adjusted to include seven day operation, base year energy costs will be correspondingly higher, and potential annual savings will be more for the same investment. Operation hours are clearly related to the cost/savings analysis of energy efficient architectural design.

As a final example, **consider adding an air-to-water heat pump on line with the electric resistance boiler** in the sample problem building. A look at the hours in each temperature bin, including **figures SP-59, Average annual HVAC energy consumption – occupied hours, and SP-64, Average annual HVAC energy consumption – unoccupied hours,** shows that most of the hours requiring heating each year happen when outdoor temperature is above 35°F. These figures show that total consumption of the electric boiler when outdoor temperature is above 30°F is 1,981,432 KWH. This is 83% of total annual boiler consumption. The boiler makes up 30% of the building's total annual electric consumption. If the heat pump is estimated to reduce boiler consumption by 66%, maximum savings of 66% of 83% of 30%, or 16% can be expected.

The same figures show the boiler required 1059 KW during unoccupied hours at 32°F. This corresponds to about 225 tons of heat pump capacity, which means that at least two heat pumps may be required. Since one will be idle when outdoor air temperatures are above 55°F, **the payback period on the first installed heat pump is much shorter than on the second. Any piece of equipment pays for itself faster if it operates for more hours at full load most of the time.** Thus, the energy planner may want to evaluate installing one heat pump only, rather than providing heat pumps with capacity to handle the entire heating load at 35°F (but operating at much less than full load most of the time).

Adding heat pumps is another example of an opportunity that has no impact on such criteria as comfort and appearance and is, thus, easier to evaluate.

Evaluating each opportunity using established criteria

Next, methodically compare the impact of every modification against each evaluation criterion. For instance, if a modification is estimated to save $100,000 the first year and cost $275,000, will it meet the costs vs. benefit criteria? Are funds available? Does it meet comfort and functional impact criteria?

In some cases, a modification could be carried out even at this early stage. If it is cost-effective and will not harm overall performance of the building, there is no point in delaying further. Examples include repairing damage to broken dampers on outdoor air intakes or reducing building operation hours.

Listing and ranking opportunities

Once each opportunity has been looked into with respect to possible savings, costs and the other criteria, three lists should be made:

Opportunities which should be carried out now as they are obviously rewarding.
Opportunities to be evaluated in more detail.
Major opportunities that failed to pass initial evaluation.

Opportunities to be evaluated in more detail should be ranked, with the most promising items listed first. With each one, record estimates and notes regarding criteria considered.

Owners should review the list so they understand the basis for all recommendations. They should then give approval as to which items should be carried out, which evaluated in greater detail, and what the final evaluation criteria should be.

Performing the detailed evaluation

Quick evaluation resembles architectural feasibility and programming. Detailed evaluation is like design development in the architectural design process. That is because costs, actual materials and details of implementation are being worked out to provide the information for an accurate evaluation.

"Good" recommendations are sorted more thoroughly and in greater detail than before. Half of a cost/benefit ratio is in costs, so **preliminary design drawings and specifications** for some items may need to be made at this time and distributed to contractors and suppliers for prices. Building operators and maintenance staff should be consulted to determine any cost changes in their areas of responsibility. If the owner requires a detailed life-cycle cost study, the owner's accountant should help predict the impact of interest, tax changes and other financial factors. These items may have been considered before; now their effect should be estimated accurately.

Samples should be obtained from manufacturers so demonstration areas can be set up. **Tests** to evaluate any questionable conditions of data input may be required. This is the time to make decisions, so anything that will help predict performance should be made use of. That includes calculations, computer simulations, tests, models, drawings, surveys of occupants, monitoring of operations and preliminary discussion with product manufacturers.

During the quick evaluation, most major factors in need of a closer look were identified. This way energy planners see where the bulk of the work lies. They now have drawn up a work schedule and set an **"in-house" budget.** The detailed evaluation refines possible energy savings items, gathers and analyzes enough information to compare each opportunity with the evaluation criteria, and accurately estimates costs involved.

The energy and cost savings estimate may take one of two forms:

"Modified" estimate: revision of the energy analysis so savings may be found by subtracting the modified estimate's results from the original totals.

Direct "differential" estimate: the difference is estimated directly. This is the simplest way if the estimate does not provide a convenient basis for comparing a particular item and if the item does not interact with other energy consuming items. For example, energy consumption of the domestic hot water heater may not have been estimated in the energy analysis, but may now need to be in order to evaluate the result of lowering supply water temperature to 90°F. The two estimates may easily be made and the difference figured, without having to rework the energy analysis, since domestic hot water will usually not influence other items.

If the first form is required for the energy and cost savings estimate, the first step is to make sure that the energy analysis was prepared in a way that will permit a true comparison: For example, a modification may be to replace poorly fitting single pane windows with new weatherstripped double-pane units. One benefit is reduced infiltration of outside air, particularly in winter. If the energy analysis overlooked infiltration, or if the energy planner merely revised the analysis results to show the improved "U" value of double glass, the energy savings would be underestimated. Infiltration would have to be incorporated as a load into the previous analysis and carried through the systems response and energy consumption steps before the impact could be figured. This modification may have a big influence on much of the building's equipment and cannot be evaluated directly, as was the case with domestic hot water.

In making a revised estimate, **the revision must start at the first point where the modification affects the estimate.** A change in equipment can reuse all calculations including those for loads and systems response, whereas an architectural change could affect building thermal loads, the systems response, as well as equipment energy consumption. This would call for complete recalculations of the analysis. However, the reduction of loads often leads to great savings.

If some modifications have been already accepted, the revised estimate must first incorporate these changes because energy-related modifications are not cumulative. For example, if a fan is modified so it consumes 10% less energy, and the fan's operating hours are reduced 10%, total fan savings are 10% plus 9% (not 10% plus 10%) since 10% of 90% = 9%.

If the energy planner needs to rework the energy analysis before proceeding with the detailed evaluation, there are two practical approaches.

First, use the same basic techniques as before, but with increasing detail, comprehensiveness and accuracy, so that more opportunities may be evaluated. Changes would also include the impact of any opportunities already slated to be carried through.

Second, rework the estimate using a different technique. If a manual estimating method was used before to identify the opportunities, a computer simulation might now be used to prove them. It could simulate an aspect of performance shown during the quick evaluation to be critical or simulate the impact of the sun or the effect of the building's mass on air-conditioning costs, for example. The owner may be interested in so many possible improvements that the additional effort in redoing the original estimate is balanced by the new method and the time it saves.

One must **plan which combinations of potential modifications are to be analyzed.** Modifications which the owner wants to implement now must go into the analysis. In addition, some modifications may be so interrelated that they must be analyzed jointly to see the final effect. Which ones to group depends on these interrelationships, the likelihood of their being carried through, and the resources available for testing combinations.

5 Evaluate Opportunities

Implementation & Followthrough

Carrying out the energy planner's recommendations, including proper operation, maintenance and monitoring, will realize the fruits of the energy planning project, namely, improved building performance and lower operating costs.

Clear and complete user's and operator's manuals are a key to continued energy efficiency. Manuals should spell out what should be done, when and how, and provide the information for it.

The owner should arrange for user and operator briefings and arrange periodic inspections to insure that the manuals are understood and used.

Once recommended modifications are carried out, it is crucial to:

Check to make sure that work has been installed as specified.

Maintain items according to a regular schedule for best performance.

Monitor performance to detect problems.

Introduction

Implementation and followthrough for an energy planning project are not unlike standard renovation work. What follows stresses those aspects that are unique to energy planning.

Implementation

Compensation for carrying out an energy planning project varies, as it did during previous phases, since it includes work not tied to construction costs but to level of service (such as preparation of operation and maintenance manuals). Also, owner and staff may put in many of the modifications themselves.

Project scheduling also is affected. The owner may wish to carry through all the energy planner's recommendations at once to gain maximum benefit from improved performance, or prefer to phase the work.

Phasing is not unusual on a construction project. It may apply especially to energy planning, since such projects commonly are linked more to investment for a return than are construction projects. The owner may wish to take all or part of the project's costs out of the operations budget. This may call for phasing the project over a few years and doing the most promising modifications first. Also operations and maintenance staff may gradually incorporate some of the work into their normal procedures.

Another use of phasing is for the building owner who cannot increase the energy supply, but wants to expand a building complex. Energy performance of existing buildings may need to be improved to free energy for use in new buildings or additions. Part of the costs may come from the expansion budget. In some cases, building owners may have oversized HVAC equipment which could also condition building additions and **save on initial HVAC costs.** The work to reduce HVAC loads on existing buildings will be a credit toward the new additions.

Organizing the team

The various team members must be coordinated during implementation and followthrough. **Users, operations and maintenance staff must be kept informed on progress;** they must be taught how to operate and maintain the building; and their schedules must be coordinated with contractors' and suppliers' work schedules.

Organizing the team is up to the owner. The energy planner helps in planning the work schedule and supplies necessary information. Owners may, if they wish, turn over more of their responsibilities to the energy planner.

An information program may need to be devised for building users. This will help them understand why various measures are being taken, gain their support and instruct them as to new use procedures. If a user's manual has been prepared, it should be explained and passed out. Information programs can range from memos to full-scale "mock-ups." Owners may only want to instruct tenants or department heads, and leave it to them to pass information on to other users. Or they may want to include all who use the building.

Owners may wish to retain the energy planner to conduct the information program. Since the way people use the building is crucial to how it performs, **information programs should be thought of not as an expense, but as an opportunity for generating real savings.**

Likewise, those who operate and maintain the building and its equipment must be kept informed. Greater technical detail will be required. Since most projects are done on a multiple contract basis, the owner or energy planner should organize contractors before work begins and make sure their schedules are compatible with day-to-day use of the building. Contractors experienced in work on occupied buildings can be very valuable. They should be included on the team as construction schedules are planned to ensure an efficient project.

On some projects, **many of the items may be done on a portion of the building first, and the impact assessed before proceeding.** Special arrangements must be made with the contractors and users involved; it may call for a separate contract for this portion of the work.

Because a building's energy performance and users' comfort are so interrelated, many items may need to be carried out one at a time before final results can be appreciated. For example, doing a more efficient lighting design may first call for installing some form of window glare control, changing some interior finishes and reorganizing the furniture or partitioning system. Next, the new lighting system can be installed. HVAC equipment can then be rebalanced and air quantities reduced to take advantage of the reduced loads.

Preparing contract documents

The energy planner will assume a more familiar architect's or engineer's role as he or she completes designs and prepares contract documents for the work. There are some differences, however.

In preparing actual contract documents, the energy planner will be relying more heavily on specifications. Many of the changes to the building, such as replacing windows, require standard working drawings. Others may require none, such as changing a fan's operation from one to two speeds. For operational improvements, the operations and maintenance manual will be the final "product," no drawings being involved.

Preparing user's and operator's manuals

Written manuals or directions remind all to continue to use and operate the building properly after modifications are in place.

How thorough each must be depends on how crucial each group's actions are to the building's performance. For instance, if the building's HVAC equipment and lighting fixtures are centrally controlled, and there are no manually controlled architectural fixtures such as sun control devices, users may have little day-to-day influence on energy performance. The user's manual in that case simply explains reasons behind the building's operation hours, room temperatures, etc., and a memo may suffice.

On the other hand, if the building has architectural fixtures and equipment designed to be controlled by the building's users, the user's manual will need to explain how each feature should be used for greatest comfort and energy performance, and why. This way, the user will appreciate each feature and be more interested in using it properly. Examples are light switches, operable windows and sun control devices. If these items are not used properly, they may worsen the situation.

Similarly, the level of detail of the operator's and maintenance crew's manual will reflect the degree of influence each has on energy performance.

A thorough operation and maintenance manual will include operating instructions for the building and its equipment, a detailed maintenance schedule and information for maintenance and repairs. The manual will reflect operation and maintenance assumptions upon which the energy planner's performance studies and resulting recommendations are based. The building's operators must know this plan and the reasoning behind it.

The manual's operations instruction should include:

- The daily schedule for opening and closing the building and special use areas within it.
- Daily times and sequences for starting up and shutting down equipment and lighting fixtures.
- Seasonal start-up, shutdown and changeover procedures.
- Control set points.

The energy planner may also include specific instructions for either load shaving or peak load shedding. Peak load shedding schedules list the priority and sequence for shutting down and starting up equipment to control the building's peak energy demand as it reaches maximum demand level. If automatic controls are used to maintain any such schedules, the manual should include enough about how they should function so building operators may easily detect problems.

A maintenance schedule should be included. Good maintenance of the building itself and its equipment makes for energy efficiency. The schedule covers periodic inspection of all important parts, as well as performance of regular tasks needed for continued efficiency and long service life, such as cleaning, lubricating, filter replacement, lamp replacement and packing of bearings. Indications of malfunctions, such as leaks, excessive vibration, corroded or worn parts, must be corrected promptly.

A list of important maintenance tasks to be performed on a regular schedule is included in the appendix. The list includes periodic inspection of all controls, as many will need occasional recalibration.

Finally, the manual should include **information required to make repairs.** A good manual will contain for the building and each piece of equipment:

- Construction drawings and specifications
- As-built and shop drawings
- Warranties and submittal data
- Purchase orders
- Original performance reports (such

as the test and balance report)
Manufacturer's installation, service and maintenance instructions, including troubleshooting checklists and spare parts lists
Lists of service contractors and spare part sources
System diagrams and control diagrams
Illustration of operating modes, especially in case of solar heating or heat storage systems

A complete manual may fill one or more large notebooks. Often, three or more copies of each are made. One should be filed in the main equipment room where it is easily found by service personnel; one should be kept by the owner; and one by outside service contractors.

A complete manual probably includes many items not directly related to energy or the energy planner's concern. For instance, it may specify what detergents can be used on carpets. These will not concern the energy planner. On the other hand, whether all lights are to be left on as a result of cleaning procedures, or only lights on the floor being cleaned, matters a great deal. Consequently, the energy planner will be responsible for those aspects of the building's performance having to do with energy, and leave the rest of the manual to others.

The owner must make sure the manual is used. If it is not specific about what must be done and when, chances are nothing will be done until there is a breakdown. By then, the building will have been operating inefficiently for some time. If procedures are clear, however, they are likely to be followed, and the building's performance maintained.

Personnel also should realize the importance of operation and maintenance procedures. **There is a self-defeating tendency to let maintenance and repairs slide in the interest of keeping operating costs down.** The crew should know that what is being called for in the manual is crucial to keeping operating costs at a minimum, and be impressed with the importance of prompt action.

By the same token, the owner or management staff must understand and approve work hours and direct expenses for carrying out the maintenance program. That way the maintenance staff is not forced to cut back due to not enough funds.

The building's equipment, distribution systems and controls must be clearly labeled so operations and maintenance staff can easily find the relationship of controls to zones, etc. Even a comprehensive manual will be hard to use if key controls, thermostats, gauges and dampers are not easy to find or get at.

Followthrough

Once the work is in place and the user's and operator's manuals prepared, three important things must be done to insure continued energy efficiency.

Check performance to make sure all is installed as specified
Maintain everything regularly to keep up performance
Monitor performance to detect problems.

Checking performance

A thorough performance check must be made upon completion of the work. The work must satisfy contract documents, and perform as expected. This is straightforward for many of the recommendations and can be done by the energy planner himself or a designee of the owner.

If any major changes were made in the HVAC system, a certified HVAC test and balance specialist will be required. Testing and balancing were no doubt recommended by the energy planner as an important part of the work, so this is not an unexpected expense. There are currently two certifying agencies. The Associated Air Balance Council (AABC) certifies independent contractors who do no installation work themselves. The National Environmental Balancing Bureau (NEBB) certifies both independents and installation contractors.

It is not unusual for an installation contractor who is also certified to test and balance to test his own work. The NEBB has a detailed set of procedures such a contractor must follow for all HVAC systems, and standard sheets on which to record test results and affix his seal.

Energy planners should review these procedures for themselves to see if there are any added tests they would like performed. The outside testing and balancing contractor should be included on the team early so any recommendations can be incorporated into the design, and to avoid post-construction charges.

Maintenance

Features added to the building to conserve energy may, if not maintained properly, end up causing the building to consume not less but more energy. Sun control devices which no longer operate may block out beneficial sunlight during the heating season, or broken door closers may keep doors ajar all day.

The building's regular staff can do many of these tasks, but some maintenance chores, such as calibrating controls, may call for an outside contractor. Even such simple items as checking a boiler's flame may be beyond the expertise of most maintenance crews. The energy planner can help the owner in staffing a maintenance program to meet the building's specific needs.

Monitoring

Finally, the building's performance should be monitored on a regular basis, and a thorough periodic check made of the building and its equipment. For best performance, problems must be detected and eliminated promptly.

The most basic items to monitor are the utility bills themselves. Each month's electric and fuel costs should be recorded, along with the consumption and demand figures. If any problem affecting a building's energy performance develops, it will show up in the utility bills. Recording meters, such as the General Electric G-9 demand meter referred to in Chapter 3, provide information as to actual time of energy use as well as the amount.

For more detail, the major pieces of equipment such as chillers, boilers and lights can be sub-metered to record actual energy use. The building's operator may want to read all meters personally several times each month so as to identify problems promptly.

Monitoring energy use in this way will allow the building operator to compare any unusual patterns (as evident from bills) with operations and maintenance records, and with weather data for the period.

Keeping records is an important part of any monitoring program. As a minimum, daily records should be made up of when the building is opened and closed, when equipment is started up and shut down, what maintenance procedures were carried out, and weather conditions, such as outside wet bulb and dry bulb temperatures, wind velocity and direction, cloud coverage and precipitation. Complaints from users should be recorded as these may also point to location and cause of any unusual energy use.

If a problem cannot be traced to operations, maintenance or weather, some part of the building or its equipment may be at fault. The next useful level of detail for locating problems, then, is to monitor comfort conditions in rooms and operating conditions of equipment. Room temperatures and humidities can be easily recorded during daily maintenance rounds, and thermometers and pressure gauges can be installed on equipment so readings can be made of incoming and outgoing temperatures and pressure.

Thus, it is best if operators can read incoming and outgoing temperatures on condensers, chillers, heat exchangers and boilers; incoming and outgoing water pressures through coils, chillers or condensers; suction and discharge on all pumps; and incoming and outgoing air temperatures and pressures on all air handling units. If an unusually large pressure drop is observed through a coil, it should be checked for obstructions or dirt. Other unusual readings will indicate problems in need of prompt attention. They will be easier to spot if gauges and thermometers are provided with a warning band, or if normal temperature and pressure ranges are clearly marked on the equipment, or both.

Many newer buildings come with elaborate central control rooms. These have all the gauges for the building's comfort conditions and equipment performance in one place, so it is easy to record performance hourly. In older buildings this may have to be done manually, but **even once-a-day checks can prevent costly down time and inefficient performance.** Many manufacturers now offer built-in monitoring devices on their equipment.

As with controls, calibration of monitoring devices should be checked periodically. Inaccurate readings are not useful.

Aside from regular in-house monitoring, **a thorough check of the various systems should be made at one-half or one year intervals by someone not involved with daily monitoring and maintenance, say a supervisor or outside service contractor.** They should thoroughly check the building and its equipment performance, evaluate its state of repair and make recommendations for any service required. This should be detailed enough to assess performance of the building and its major equipment, though less detailed than the original testing and balancing.

The **owner or manager may want to review all the monitoring records periodically,** especially utility bills, comfort conditions and complaint records. This will usually be enough to spot problems. Additional information will help staff locate the cause of problems and correct the matter.

Monitoring should also include a record of changes in building occupancy, building use, addition of office or other equipment and changes in comfort requirements as types and needs of occupants change.

The energy planning project began with an assessment of the building's performance. This process of performance assessment must continue by means of a planned monitoring program, so problems can be avoided and to obtain a basis for evaluating the effect of modifications already made and opportunities yet to be identified.

The energy planning project may have been completed, but energy planning as a process goes on, as conditions change and new opportunities arise.

Sample Problem

Problem:
Estimate the present annual energy costs for a hypothetical twenty story, relatively energy-efficient, all electric office building in Atlanta, Georgia.

Solution:
The procedure illustrated follows the format described in Chapter 3 and outlined in the Table of Contents.

All necessary data are given as they might be organized for the energy analysis.

The "modified bin method" is used for the HVAC energy estimate. Examples of both the block load calculation format and a zone-by-zone, hour-by-hour format are given with complete system response calculations.

The equivalent full-load hours method is illustrated in the non-HVAC energy calculations. Each of the major components of non-HVAC equipment is discussed separately.

Calculations show that the building consumes an average of 8,076,000

Introduction

The intent of the sample problem is to let interested readers carry out actual calculations for an energy analysis using the authors' "modified bin method," by following the steps involved and trying to duplicate them if they wish. The problem has been kept simple and straightforward: the aim is to explain method and process rather than details. Every building is unique. So the example is complicated enough to show up all the procedures, yet simple enough to be followed by most readers.

The sample problem building is a hypothetical 20-story office structure located in Atlanta, Ga. It is in part a composite of elements from existing structures.

The solution is arranged to correspond to the discussion of an energy analysis in Chapter 3. If the rationale for a particular step appears obscure, refer to Chapter 3 for more explanation.

*One need only understand how to do the analysis of the building **in its present state** to be able to analyze proposed **modifications.** That is so because evaluating such modifications calls for re-doing all or part of the energy analysis to assess the impact of any changes. The sample problem can thus be used to illustrate how to identify system modifications cited in Chapter 4 and how to evaluate them in Chapter 5.*

*Those using this chapter are assumed to be familiar with normal HVAC design calculation procedures, and interested chiefly in the type and sequence of calculations. Those without HVAC experience should consult a good basic work, such as **The ASHRAE Cooling and Heating Load Calculation Manual.***

More experienced readers should continue to use their present data source. Any slight differences between manuals will not inhibit understanding of this format. The authors used the Carrier System Design Manual.

The Building: First Floor

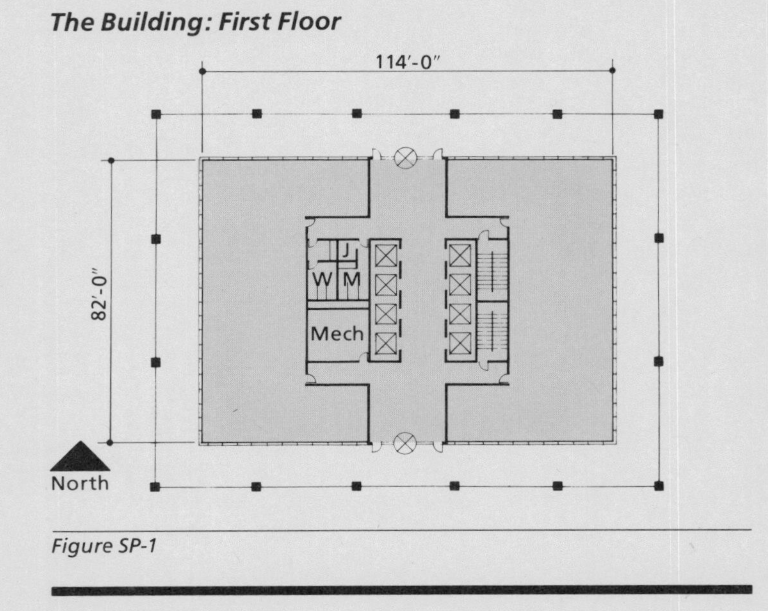

Figure SP-1

The Building: Floors 2-20

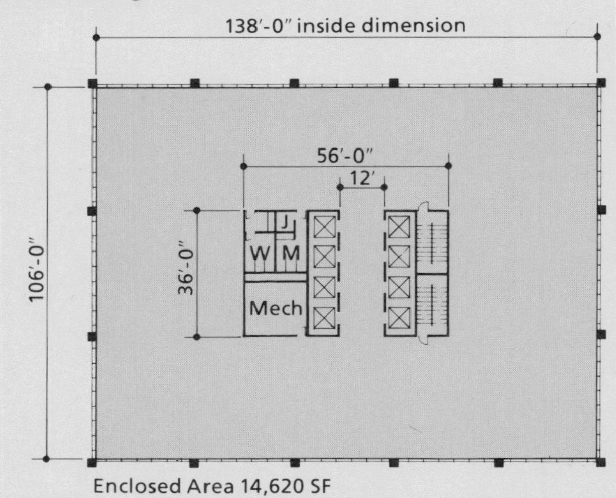

Figure SP-2

KWH per year which amounts to 26 KWH/gross sq. ft. or 89 MBTU/gross sq. ft. annually. The building's peak energy demand is 3567 KW and occurs in the winter months when the electric boiler is running at capacity.

Applying the local Georgia Power rate schedule to the calculated monthly peak energy demand and average energy consumption figures gives a present annual energy cost of $300,577.64. Because the rate structure penalizes summer peaking, the demand of 2518 KW set in September determines the monthly bills rather than the higher winter peak demand.

Since the calculation method shows all the loads on the building's equipment and the equipment's response to these loads, it is possible to quickly identify major opportunities for improvement and evaluate their effectiveness. Graphic representations of the calculations are included to help in this regard following the annual energy use and cost summaries.

Data

The data displayed here and in the sample problem figures includes only what is needed to work the problem. The extent of needed data will vary from actual project to project.

Data

Building

Gross areas

A. First floor	114′ x 82′ =	9,348 sq. ft.
B. Floors 2-20	@138′ x 106′ =	277,932 sq. ft.
C. Equipment room	138′ x 106′ =	14,628 sq. ft.
D. 3 levels parking (fl. to fl. ht., 9′)	@176′ x 142′ =	74,976 sq. ft.
Total gross area	A + B + ¼ C + ¼ D =	309,681 sq. ft.

Conditioned areas

First floor	7,764 sq. ft.
Floors 2-20	247,836 sq. ft.
	255,600 sq. ft.

Structure
Steel frame with light-weight joists and metal deck.

Exterior opaque walls
6″ precast concrete panels, exposed aggregate with 1″ rigid insulation inside.

Glass

First floor	¼″ fixed clear glass
Floors 2-20	¼″ fixed gray heat absorbing glass

Shading devices

Floors 2-20	inside venetian blinds, white

Floors

Equipment room	8″ concrete – no insulation
Floors 1-20	8″ concrete with carpet

Roof
Built-up roofing over 2″ of rigid insulation on metal deck.

Use
The building is used for general offices, five days per week. It is kept open from 7 AM until 10 PM for the convenience of tenants who arrive early or need to stay late. Assume 51 weeks of use per year to account for holidays. HVAC system operation is from 6 AM through 10 PM **(figs. SP1,2,3,4,5,6).**

The Building: Section

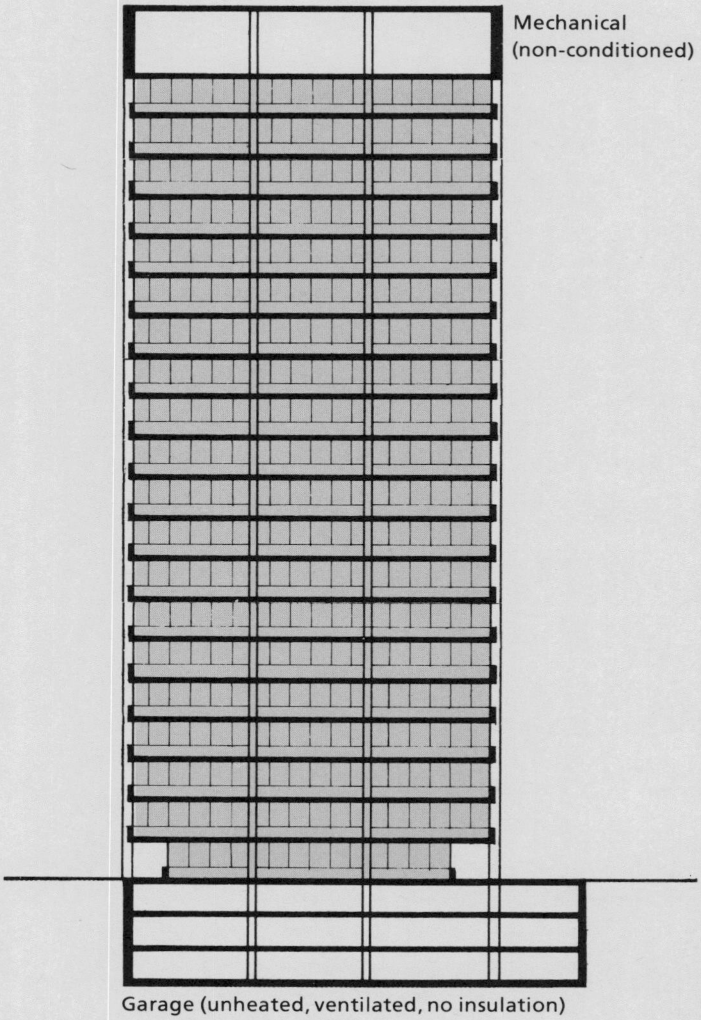

Mechanical (non-conditioned)

Garage (unheated, ventilated, no insulation)

Figure SP-3

Sample Problem

The Building: Wall Section

Section

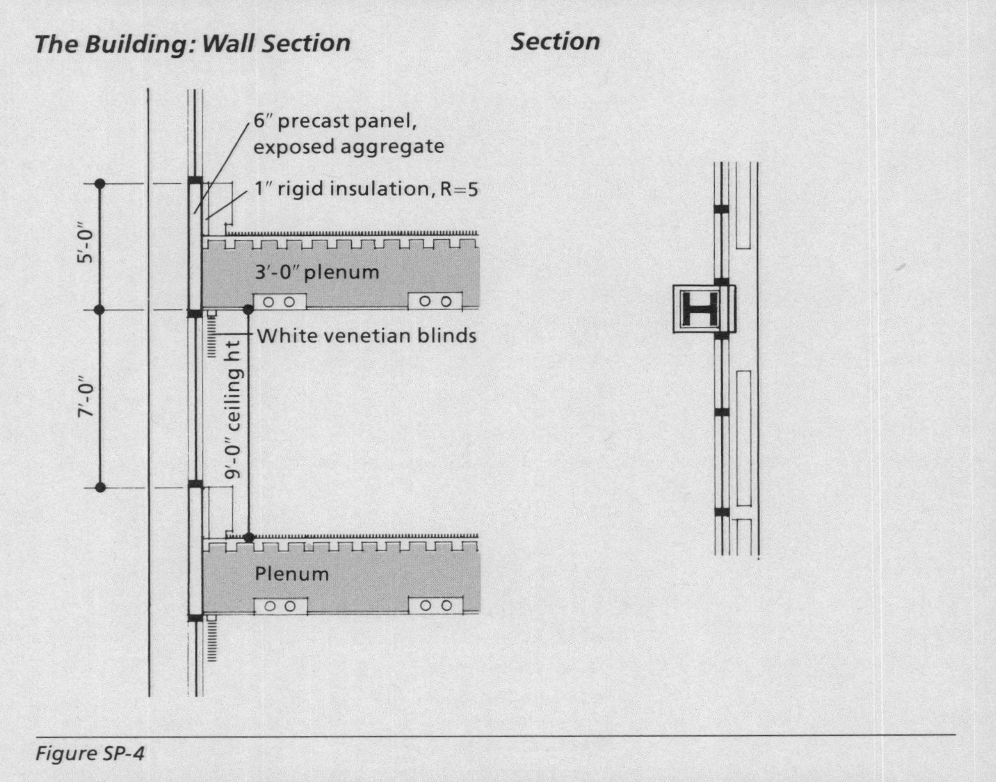

Figure SP-4

The Building: HVAC Zones

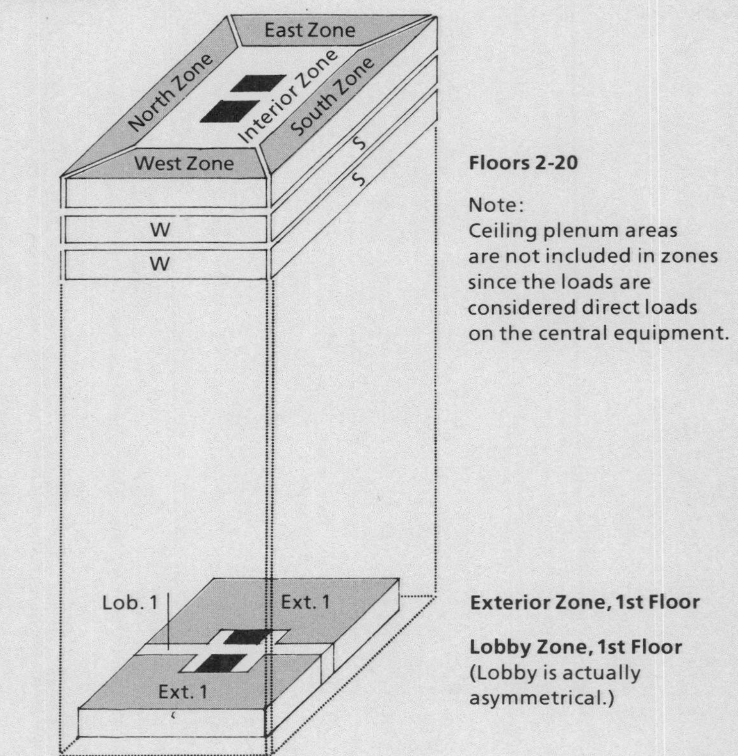

Figure SP-5

Throughout the figures in the Sample Problem certain data is printed in green. This indicates data gathered, calculated or derived by the energy planner during the course of his services.

Building Data by Zone

Zone		N	E	S	W	Int.	Ext. 1	Lob. 1	Total
Floor Area (sq. ft.)		28,728	21,432	28,728	21,432	147,516	5776	1988	255,600
Envelope Materials									
Floors	Area (sq. ft.)						5776	1988	
	U-Value						.32	.32	
Roof	Area (sq. ft.)	1,512	1,128	1,512	1,128	7,764			
	U-Value	.15	.15	.15	.15	.15			
Glass	Area (sq. ft.)	17,290	13,566	17,290	13,566		3060	468	
	U-Value	1.1	1.1	1.1	1.1		1.1	1.1	
	Shade Coeff.	.53	.53	.53	.53		.94	.94	
Walls									
Room Wall	Area (sq. ft.)	6308	4560	6308	4560				
	U-Value	.16	.16	.16	.16				
	Weight	Med.	Med.	Med.	Med.				
Plenum Wall	Area (sq. ft.)	7866	6042	7866	6042		1020	156	
	U-Value	.16	.16	.16	.16		.16	.16	
	Weight	Med.	Med.	Med.	Med.		Med.	Med.	
People (office workers – 75°F Room dry-bulb)									
Maximum Number*		359	268	359	268	1582	72	25	2933
Heat Gain Sensible $\left(\frac{\text{BTUH}}{\text{person}}\right)$		245	245	245	245	245	245	245	
Heat Gain Latent $\left(\frac{\text{BTUH}}{\text{person}}\right)$		205	205	205	205	205	205	205	
Office Equipment (watts)		14,364	10,716	14,364	10,716	63,270	2888		116,318

*Use 80% for Central Equipment Load (part B); 60% for Average Load (part C)

Figure SP-6

Data

Environment

The site

The building is located in Atlanta, Ga.; altitude is 975 ft., closest latitude 34°N. It is typical of many speculative offices in the area in that it stands alone, with no major surrounding structures or land forms. Atlanta has an average 2985 heating degree days, according to the *Carrier System Design Manual*.

Data

Weather

Air temperatures and humidity

Average outdoor air dry-bulb and wet-bulb temperatures are obtained from U.S. Air Force Manual 88-8, Chapter 6, "Engineering Weather Data," **(fig. SP7a,b).**

Sun

Percent sunshine figures were obtained from ASHRAE Transactions 1974, Vol. 80, Part II. Solar heat gain factors were obtained from Carrier System Design Manual.

Data

Wind

Since the building's glass is fixed, a summer infiltration rate of 0 has been assumed. For winter, the peak design computations are based on one air change per hour. (Infiltration depends on construction quality, detailing, building height and internal building air pressures. These are not easy to quantify and should be one of the first assumptions to be questioned. See also discussion on the building shell, Chapter 1, Introductory Concepts.)

Dobbins AFB Georgia

Mean Frequency of Occurrence of Dry Bulb Temperature (°F) With Mean Coincident Wet Bulb Temperature (°F) For Each Dry Bulb Temperature Range

Cooling Season

Temperature Range (°F)	May					June					July				
	Obsn. Hour Gp. 08 to 09	10 to 17	18 to 01	Total Obsn.	Mean Coincident Wet Bulb (°F)	Obsn. Hour Gp. 08 to 09	10 to 17	18 to 01	Total Obsn.	Mean Coincident Wet Bulb (°F)	Obsn. Hour Gp. 08 to 09	10 to 17	18 to 01	Total Obsn.	Mean Coincident Wet Bulb (°F)
95/99							2	0	2	74		1		1	76
90/94		4	0	4	69		20	4	24	73		24	5	29	74
85/89		36	7	43	69	1	57	13	71	72	0	79	21	100	73
80/84	2	65	20	37	67	7	71	32	110	70	3	34	43	135	72
75/79	3	60	37	106	66	27	49	57	133	69	43	44	72	169	71
70/74	27	41	53	126	64	70	25	30	175	68	125	14	90	229	69
65/69	76	22	66	164	62	91	13	45	149	65	62	2	16	80	66
60/64	60	13	29	108	58	36	2	7	44	60	10		1	11	61
55/59	31	5	20	57	54	8	1	2	11	55	0			0	56
50/54	19	2	8	29	49	1		0	1	49					
45/49	11	0	3	14	44	0			0	45					
40/44	6		0	6	41										
35/39	0			0	33										
30/34															
25/29															

Temperature Range (°F)	August					September					October				
	Obsn. Hour Gp. 08 to 09	10 to 17	18 to 01	Total Obsn.	Mean Coincident Wet Bulb (°F)	Obsn. Hour Gp. 08 to 09	10 to 17	18 to 01	Total Obsn.	Mean Coincident Wet Bulb (°F)	Obsn. Hour Gp. 08 to 09	10 to 17	18 to 01	Total Obsn.	Mean Coincident Wet Bulb (°F)
95/99		4	0	4	74		1	0	1	69					
90/94		34	5	39	74		12	1	13	71					
85/89	0	91	21	112	73		45	6	51	71		1		1	72
80/84	9	74	42	125	72	1	62	20	83	70		21	1	22	68
75/79	34	31	81	146	71	11	52	41	104	68		44	6	50	64
70/74	121	12	77	210	69	60	37	76	173	67	7	53	23	88	63
65/69	65	2	20	87	65	73	17	48	138	64	23	49	43	115	60
60/64	18		2	20	61	52	11	31	94	59	44	35	55	134	57
55/59	1			1	56	27	2	12	42	55	51	52	49	125	53
50/54						13	1	3	17	50	48	11	38	97	48
45/49						3	0	1	4	47	37	8	23	63	44
40/44						0			0	42	22	1	7	30	40
35/39											12		3	16	35
30/34											8		6	3	31
25/29											1			1	23

Figure SP-7a

Dobbins AFB Georgia

Mean Frequency of Occurrence of Dry Bulb Temperature (°F) With Mean Coincident Wet Bulb Temperature (°F) For Each Dry Bulb Temperature Range

Heating Season

Temperature Range (°F)	November					December					January				
	Obsn. Hour Gp. 08 to 09	10 to 17	18 to 01	Total Obsn.	Mean Coincident Wet Bulb (°F)	Obsn. Hour Gp. 08 to 09	10 to 17	18 to 01	Total Obsn.	Mean Coincident Wet Bulb (°F)	Obsn. Hour Gp. 08 to 09	10 to 17	18 to 01	Total Obsn.	Mean Coincident Wet Bulb (°F)
95/99															
90/94															
85/89															
80/84		1		1	67										
75/79		6	9	6	64										
70/74	0	23	2	23	61		3		3	60		1	0	1	64
65/69	7	28	14	59	69	1	16	0	17	68	2	9	2	13	60
60/64	16	41	28	88	56	7	21	12	40	57	4	15	9	28	56
55/59	26	40	40	106	52	12	28	18	58	53	9	23	12	44	62
50/54	32	37	39	108	47	16	40	28	84	47	11	30	23	70	46
45/49	39	23	44	108	43	27	41	38	106	43	22	42	32	96	43
40/44	34	15	35	84	38	39	44	46	129	38	32	48	40	116	38
35/39	41	8	20	69	34	41	27	46	114	34	40	36	49	126	34
30/34	31	3	12	46	30	45	17	33	95	29	48	23	41	112	30
25/29	9	0	5	14	25	32	7	13	57	25	35	10	20	65	25
20/24	4		1	5	21	18	2	5	26	20	26	7	11	44	20
15/19	1			1	16	5	1	2	8	16	10	2	6	18	16
10/15						3	1	1	5	11	7	1	2	10	11
5/9						1	0	1	2	6	1	0	1	2	7
0/4						1		0	1	1	0		0	0	0
−5/−1						0			0	−1	1			1	−4

Temperature Range (°F)	February					March					April				
	Obsn. Hour Gp. 08 to 09	10 to 17	18 to 01	Total Obsn.	Mean Coincident Wet Bulb (°F)	Obsn. Hour Gp. 08 to 09	10 to 17	18 to 01	Total Obsn.	Mean Coincident Wet Bulb (°F)	Obsn. Hour Gp. 08 to 09	10 to 17	18 to 01	Total Obsn.	Mean Coincident Wet Bulb (°F)
95/99															
90/94															
85/89												3	0	3	67
80/84							1	0	1	60		21	4	25	65
75/79		1		1	65		7	2	9	61	1	33	13	47	62
70/74		8	1	9	60		14	4	18	69	4	47	26	77	60
65/69	2	16	7	23	53	3	26	12	41	50	18	42	42	102	58
60/64	6	22	13	41	55	9	40	24	73	53	44	39	52	135	56
55/59	17	29	30	76	52	17	43	37	97	50	49	26	42	117	52
50/54	21	33	28	82	47	31	38	47	116	47	38	19	25	82	47
45/49	30	33	34	97	43	41	36	44	121	43	37	8	24	69	43
40/44	34	32	35	101	38	50	20	37	107	38	28	2	10	40	39
35/39	36	27	36	99	34	46	13	28	82	34	17		2	19	34
30/34	36	14	22	72	30	35	6	12	53	30	4			4	31
25/29	21	6	11	38	24	11	4	6	20	26	0			0	28
20/24	14	2	4	20	20	3	0	1	4	21					
15/19	4	1	1	6	16	1	0	0	1	17					
10/15	1	0	1	2	11	1			1	11					
5/9	2	0	1	3	6										
0/4	0			0	2										
−5/−1															

Figure SP-7b

Annual (Total – All Months) Obsn. Hour Gp. 08 to 09	10 to 17	18 to 01	Total Obsn.	Mean Coincident Wet Bulb (°F)
	8	1	9	74
	95	14	108	73
1	314	69	384	71
26	399	161	586	63
126	327	300	761	66
417	293	437	1137	64
420	233	314	937	61
310	240	262	812	57
245	221	261	730	53
229	215	240	684	48
247	191	243	631	44
244	138	213	615	39
235	110	180	525	35
204	62	120	386	30
109	27	60	196	26
66	11	22	99	20
21	4	9	34	16
12	2	3	17	11
4	0	2	6	6
1		0	1	1
1			1	−3

5°F Temperature Bin Breakdown : Sample Month: April

	(1)	(2)	(3)	(4)	(5)	(6)	(7)	(8)
Hours of Operation	**Total Hours**	**6AM-10AM**	**10AM-6PM**	**6PM-10PM**				
Hours from Air Force Bins		**(½)2 thru 9**	**10 thru 17**	**(½)18 thru 01**	**(2)+(3)+(4) Day Hours**	**$(5) \times 5/7 \times 51/52$ Normal Operating Hours**	**(5)−(6) Week-end Hours**	**(1)−(5) Night Heating Hours**
O.A. Temp. Bin								
92°								
87°	3		3		3	2	1	
82°	25		21	2	23	16	7	2
77°	47	1	33	6	40	28	12	7
72°	77	2	47	13	62	43	19	15
67°	102	9	42	21	72	50	22	30
62°	135	22	39	26	87	61	26	48
57°	117	25	26	21	72	50	22	45
52°	82	19	19	12	50	35	15	32
47°	69	19	8	12	39	27	12	30
42°	40	14	2	5	21	15	6	19
37°	19	9		1	10	7	3	9
32°	4	2			2	1	1	2
Totals	**720**					**335**	**146**	**239**
Additional Months								
Annual Totals						**4092**	**1752**	**2916**

Figure SP-8

Data

The preceding temperature bin data from AF Manual 88-8 have been re-arranged for use in the HVAC energy estimates. In the data, hours of occurrence for each temperature bin are broken down into three segments: 2 AM through 9 AM; 10 AM through 5 PM; and 6 PM through 1 AM. This breakdown must be interpolated so it corresponds to the building's HVAC operation schedule, which is 6 AM –10 PM, Monday to Friday, for full air-conditioning and 24 hours each day as needed for heating **(see fig. SP-8)**

Column 1 has total hours in each temperature bin for each month.

Column 2 gives total hours in each temperature bin that occurred between 6 AM and 10 AM. To obtain this, divide total hours of occurrence from the Air Force charts for each bin between 2 AM and 10 AM by two, since 6 AM to 10 AM (4 hours) is half the hours of 2 AM to 10 AM (8 hours).

Column 3 gives total hours of occurrence between 10 AM and 6 PM. This needed no interpolation.

Column 4 gives total hours of occurrence between 6 PM and 10 PM. As in column 2, interpolation was done by dividing total hours of occurrence by two, since 6 PM–10 PM is half the hours of 6 PM to 2 AM.

Column 5 is simply the total of columns 2 + 3 + 4.

Column 6, normal HVAC operation hours, was obtained by multiplying the figure in Column 5 by 5/7 to account for weekend off days, then by 51/52 to account for holidays.

Column 7, weekend hours, includes an equivalent week of holidays. The hours were calculated by subtracting normal operating hours (column 6) from total day hours (column 5).

Column 8, night heating hours, is the difference between the total number of hours in each bin (column 1) and the day hours (column 5). The building's unoccupied period is the sum of columns 7 and 8.

Sample Problem

Data

Equipment
This is an all-electric building. The HVAC system is a VAV (variable air volume) interior system set at 60°F supply air with a four-pipe fan coil system. Central equipment includes a water-cooled refrigeration system and an electric resistance boiler. Domestic hot water is provided by an independent electric resistance heater. Fluorescent lighting fixtures are used in all tenant areas and in the parking garage. Other areas use incandescent lighting. There is no significant process equipment so a general allowance for installed office equipment of 0.5 watts/sq. ft. was used. Inventories are listed in **figs. SP-9, 10, 11.**

Central Equipment Inventory

Item	Ident.	Size	Voltage Nominal	(Volts) Actual	Amperage Nominal	(Amps) Actual	Installed Capacity	Information Source	Notes
H.V.A.C.									
North fan coil units	FC-1	—	115	—	—	—	4,200 BTUH per unit	Manu	80w Fan Motor
East fan coil units	FC-2	—	115	—	—	—	8,300 BTUH per unit	Manu	170w Fan Motor
South fan coil units	FC-3	—	115	—	—	—	8,300 BTUH per unit	Manu	170w Fan Motor
West fan coil units	FC-4	—	115	—	—	—	8,300 BTUH per unit	Manu	170w Fan Motor
1st. Fl. office fan coil units	FC-5	—	115	—	—	—	26,000 BTUH per unit	Manu	466w Fan Motor
1st. Fl. lobby fan coil units	FC-6	3/4 HP	460	467	1.3	1.2	32,000 BTUH per unit	Manu	
V.A.V. supply fan	S-1	200 HP	460	467	240	varies	163,000 CFM @ 5.5" H_2O	Plans	
Ch. water pump Total: 2	P-1, P-2	60 HP	460	467	73.5	69.9	2160 GPM @ 80' H_2O	Plate	
Hot water pump Total: 2	P-3, P-4	15 HP	460	467	20.6	15.9	606 GPM @ 60' H_2O	Plate	
Cond. water pump Total: 2	P-5, P-6	50HP	460	467	61.3	52.6	2700 GPM @ 50' H_2O	Plate	
Water chiller	CH-1	900 tons	460	467	—	—	900 tons @ 44° Ch. Wtr. @ 95° Cond. Wtr.	Plans	670 KW input Supply air temp 60°
Cooling tower	CT-1	6-15 HP	460	467	19.7	18.8	2700 GPM 95° to 85° @ 78° W.B.	Plans	
Boiler	EB-1	2000 KW	460	467	—	—	606 GPM 157.5° to 180°	Plans	
Toilet exhaust fan	E-1	7½ HP	460	467	10.3	9.3	21,000 CFM @ 1.0" H_2O	Plans	
Equipment room Exhaust fan	E-2	10 HP	460	467	13.8	10.7	30,000 CFM @ .375" H_2O	Plans	
Garage exhaust fan Total: 3	E3, E4, E5	5 HP	460	467	6.9	5.2	15,000 CFM @ .375" H_2O	Plans	
Elevators	**Speed**								
4 @ 3500 lbs.	700 FPM	40 HP	460	467	varies	varies			
4 @ 3500 lbs.	500 FPM	30 HP	460	467	varies	varies			
Other Equipment									
Water heater	EB-2	100 KW	460	467	—	—	430 GPM @ 40°-140° F. Rise	Plans	
House pumps Total: 2	P7-P8	7½ HP	460	467	10.3	7.1	125 GPM @ 120 ft. head	Plate	

Manu: Manufacturer
Plans: Plans and specs
Plate: Name plate

Figure SP-9

Zone Inventories

Zone Lighting Inventory

Conditioned Areas

Type	Zone	N	E	S	W	Internal	External-1	Lobby-1	Total
2′ x 4′ recessed fluorescent (@ 200 w)	Number	575	428	575	428	2531	115		4652
	Total Watts	115,000	85,600	115,000	85,600	506,200	23,000		930,400
	heat to space	75%	75%	75%	75%	75%	75%		
	heat to plenum	25%	25%	25%	25%	25%	25%		
Recessed incandescent downlights (@ 150 w)	Number					247		29	276
	Total Watts					37,050		4,350	41,400
	heat to space					100%		100%	
	heat to plenum					—		—	

Unconditioned Areas

Type	Zone	Core	Perimeter Overhang	Parking	Equipment Room
Surface-mounted incandescent (@ 100 w)	Number	254			300
	Total Watts	25,400			30,000
Recessed incandescent (@ 150 w)	Number		34		
	Total Watts		5,100		
2′ x 8′ Fluorescent strip (@ 100 w)	Number			750	
	Total Watts			75,000	

Figure SP-10

Data

Operating methods
These are summarized in **fig. SP-11.**

The building's energy sources
In Atlanta, this all-electric building would be billed according to Georgia Power Company's Schedule "PL-1" **see figs. SP-12, 13).**

Operating Methods

Assume 51 weeks per year

Item	Hours	Air Force 88-8 Bin Hours (02 thru 09 / 10 thru 07 / 18 thru 01; 02 04 06 08 10 12 14 16 18 20 22 24)	Hours/ Week
Building Use			
Hours open	7 AM – 10 PM, M-F		75
Hours fully occupied	8 AM – 12 PM 1 PM – 5 PM M-F		40
Hours cleaned	5 PM – 10 PM, M-F		25
HVAC System Operation			
Normal schedule	6 AM – 10 PM, M-F		80
Stand-by heating mode	Nights and weekends when outside temperature is below 75°F		
Exhaust and Ventilation Equipment			
Toilet exhaust	6 AM – 10 PM, M-F		80
Garage exhaust	7 AM – 10 PM, M-F		75
Equipment room ventilation	Varies with outside temperature	Varies	
Lighting Operation			
Tenant spaces, lobby and corridor	8 AM – 6 PM, M-F 100%		50
	6 PM – 10 PM, M-F 20%		20
Core	7 AM – 10 PM, M-F 100%		75
Perimenter	Time clock set monthly		28 (avg)
Parking	7 AM – 10 PM, M-F 100%		75
Equipment room	6 AM – 10 PM, M-F 100%		80
Elevator Operation			
	7 AM – 10 PM. M-F as needed		N/A
Other Equipment Operation			
Domestic water heater	As needed	Indeterminant	N/A
House pumps	Varies – appx. 6-8 hours/day per day – M-F	Indeterminant	~34
General office equipment	8 AM – 12 PM 1 PM – 5 PM, M-F Use 50% for central equipment peak demand and use 25% for average consumption.		40

Figure SP-11

Georgia Power Company
Power and Light Schedule "PL-1"

Availability:
Throughout the Company's service area from existing lines of adequate capacity.

Applicability:
To all electric service required on Customer's premises, delivered at one point and metered at or compensated to that voltage.
All service subject to the Rules and Regulations of Electric Service on file with the Georgia Public Service Commission.

Type of Service:
Single or three phase, 60 hertz, at a standard voltage.

Monthly Rate – Energy Charge Including Demand Charge:

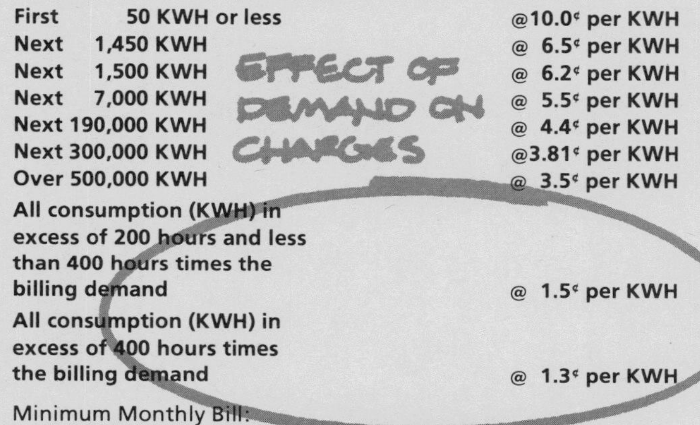

First	50 KWH or less	@10.0¢ per KWH
Next	1,450 KWH	@ 6.5¢ per KWH
Next	1,500 KWH	@ 6.2¢ per KWH
Next	7,000 KWH	@ 5.5¢ per KWH
Next	190,000 KWH	@ 4.4¢ per KWH
Next	300,000 KWH	@3.81¢ per KWH
Over	500,000 KWH	@ 3.5¢ per KWH

All consumption (KWH) in excess of 200 hours and less than 400 hours times the billing demand @ 1.5¢ per KWH

All consumption (KWH) in excess of 400 hours times the billing demand @ 1.3¢ per KWH

Minimum Monthly Bill:

A. $5.00 per meter plus $3.50 per KW of billing demand in excess of 5 KW. Plus excess KVAR charges and fuel adjustment as applied to the current month KWH.

B. Metered Outdoor Lighting: The lesser of (1) that determined from paragraph "A" above, or (2) $15.00 per meter for metered outdoor lighting installations, provided service is limited to the lighting equipment itself and such incidential load as may be required to operate coincidentally with the lighting equipment.

Fuel Adjustment:
The amount calculated at the above rate is subject to increase or decrease under the provisions of the Company's Fuel Adjustment Rider, Schedule "FA-2."

Determination of Billing Demand:
The Billing Demand shall be based on the highest 30-minute KW measurements during the current month and the preceding eleven (11) months. For the billing months of June through September, the Billing Demand shall be the greatest of (1) the current actual demand or (2) ninety-five percent (95%) of the highest actual demand occurring in any previous applicable summer month or (3) sixty percent (60%) of the highest actual demand occurring in any previous applicable winter month (October through May). For the billing months of October through May, the Billing Demand shall be the greater of (1) ninety-five percent (95%) of the highest summer month (June through September) or (2) sixty percent (60%) of the highest winter month (including the current month). In no case shall the Billing Demand be less than the contract minimum nor less than 5 KW.

Where there is an indication of a power factor of less than 90% lagging the Company may at its option, install metering equipment to measure Reactive Demand. The Reactive Demand shall be the highest 30-minute KVAR which is in excess of one-half the measured actual KW in the current month. The Company will bill excess KVAR at the rate of $0.20 per excess KVAR.

Term of Contract:
Not less than one year up to and including 500 KW maximum anticipated 30-minute KW, nor less than five years over 500 KW maximum anticipated 30 minute KW.

Figure SP-12

Georgia Power Company
Rate Schedule PL-1 (160)
Short Cut Calculation Sheet

(The short cut calculation sheet is a simplified mathematical form for calculating the base bill.) To use this chart, first divide the KWH's by the *billing* demand. (Both quantities are shown on your bill).

If the quotient is less than 200, use table (1).

If it is more than 200 but less than 400, use table (2).

If it is more than 400, use table (3).

After locating the proper chart, go to the range applicable to you and perform the calculations indicated.

Table 1

KWH Range	Correction Factor		¢/KWH
0-50	$ 0	+	10.00¢, plus F.A.
51-1500	$ 1.75	+	6.5 ¢, plus F.A.
1501-3000	$ 6.25	+	6.2 ¢, plus F.A.
3001-10000	$ 27.25	+	5.5 ¢, plus F.A.
10001-200000	$ 137.25	+	4.4 ¢, plus F.A.
200001-500000	$1,317.25	+	3.81¢, plus F.A.
500001-Over	$2,867.25	+	3.5 ¢, plus F.A.

Over 200 Hours Use — Table 2

KW Range	Correction Factor		$/KW	¢/KWH
5-7	$ 1.75	+	$10.00	+ 1.5¢, plus F.A.
8-15	$ 6.25	+	$ 9.40	+ 1.5¢, plus F.A.
16-50	$ 27.25	+	$ 8.00	+ 1.5¢, plus F.A.
51-1000	$ 137.25	+	$ 5.80	+ 1.5¢, plus F.A.
1001-2500	$1,317.25	+	$ 4.62	+ 1.5¢, plus F.A.
2501-Over	$2,867.25	+	$ 4.00	+ 1.5¢, plus F.A.

Over 400 Hours Use — Table 3

KW Range	Correction Factor		$/KW	¢/KWH
5-7	$ 1.75	+	$10.80	+ 1.3¢, plus F.A.
8-15	$ 6.25	+	$10.20	+ 1.3¢, plus F.A.
16-50	$ 27.25	+	$ 8.80	+ 1.3¢, plus F.A.
51-1000	$ 137.25	+	$ 6.60	+ 1.3¢, plus F.A.
1001-2500	$1,317.25	+	$ 5.42	+ 1.3¢, plus F.A.
2501-Over	$2,867.25	+	$ 4.80	+ 1.3¢, plus F.A.

Add sales tax and excess reactive charge of $0.20/RKVA when applicable.

Minimum Bill: $3.50 per KW of billing demand in excess of 5 KW plus $5.00 per meter + F.A. + excess reactive charge.

Figure SP-13

HVAC Energy Calculation

Part A Compare HVAC equipment capacity with requirements

Step A1 Zone cooling requirements

First calculate zone peak cooling load, system response and zone equipment selection for each exterior zone. **Figure SP-14** shows the zone peak thermal load for one example zone – in this case, the east zone, which peaks at 9 AM in August with a likely outside temperature of 83°F.

Load estimate assumptions were: 1) 25% of heat from lights and all transmission through the wall above the ceiling level goes directly into the return air ceiling plenum to the central air system. 2) The return air heat gain was not estimated because central apparatus loads are not considered in this step.

The exterior zones are actually served by two systems, so the zone system response must separate and apportion the loads to the proper equipment. Ventilation is provided by the central VAV system, which supplies 0.4 CFM per sq.ft. of air at 60°F. Zone response must reduce the zone sensible cooling load by the sensible cooling capacity of the cold ventilation air to obtain the true sensible cooling load on the fan-coil units.

The entire latent cooling load in the exterior zones is assumed to be absorbed by the dehumidified central ventilation air. Fan-coil units will operate as sensible cooling units.

Zone response calculations and required fan-coil zone equipment capacities are shown in **figure SP-15.**

Existing fan-coil units have a lot of excess capacity (8,300 BTUH capacity vs. 6,654 BTUH load), probably because the number of units was determined by the column spacing. What is more, the units come in fixed sizes – 600 CFM was too much and 400 CFM too little.

Step A1 data for all exterior zones is listed in **figure SP-16.**

The interior zone VAV system has no utility energy-consuming equipment in individual rooms, so these room design loads did not have to be verified.

Zone Equipment Capacities

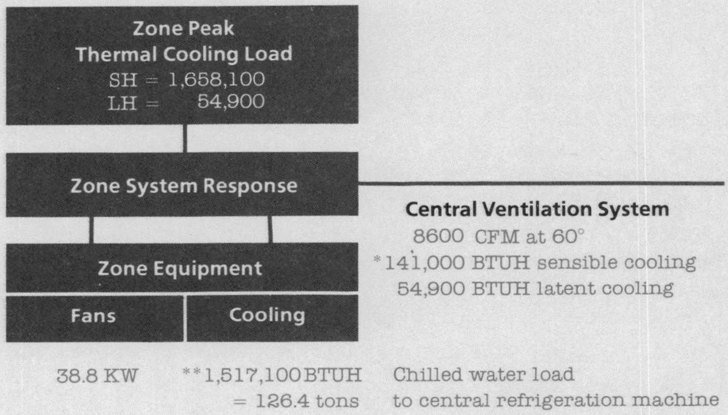

Zone Peak
Thermal Cooling Load
SH = 1,658,100
LH = 54,900

Zone System Response

Central Ventilation System
8600 CFM at 60°
*141,000 BTUH sensible cooling
54,900 BTUH latent cooling

Zone Equipment

Fans	Cooling	
38.8 KW	**1,517,100 BTUH = 126.4 tons	Chilled water load to central refrigeration machine

*Sensible cooling capacity = 8600 CFM × 1.09 × (75°-60°) = 141,000 BTUH

**Required fan-coil unit capacity:

Zone thermal load	–	Central vent system load	= Total
SH = 1,658,100	–	141,000	= 1,517,100 BTUH
LH = 54,900	–	54,900	= 0

Per Unit:
Required capacity = 1,517,100 ÷ 228 units = 6654 BTUH sensible heat
Actual capacity = 8300 BTUH sensible heat (from SP-16).

Figure SP-15

East Zone Peak Cooling Load 9am August 83°F.O.A. Temp

Line	Item	Quantity		Temp. Diff. or SHGF		"U" or S.C.	Constant BTU/KW or BTU/Person		Factors		BTUH
1	Glass solar	13566 sq. ft.	×	148	×	.53				=	1,064,100
2	Glass trans.	13566 sq. ft.	×	(83°-75°)	×	1.1				=	119,400
3	Wall trans.	4560 sq. ft.	×	26°	×	.16				=	19,000
4	Lights	86 KW	×				3413	×	75% to room	=	220,100
5	Equipment	11 KW	×				3413			=	37,500
6	People, SH.	268	×				245			=	65,700
7	Fan-coil motors	38.8 KW					3413			=	132,400
									Zone Sensible Heat	=	1,658,100
8	People, LH.	268					205		Zone Latent Heat	=	54,900

Figure SP-14

Step A2 Central system cooling requirements

This step is used to calculate the block load at a time of overall building peak load. This is to verify the central apparatus and equipment capacities. **Figure SP-17** shows the block load at 4 PM in August, which the authors estimate as peak block building load time. The block load format is used instead of calculating individual zone loads because all the fan-coil units require cooling at this time and no reheat is involved.

The first floor glass and plenum wall areas on the east, south and west are shaded by a 12 ft. overhang and receive little direct sunlight. Their cooling loads are similar to that of a north exposure. For load estimating purposes, they therefore were lumped together with the north glass (line 1) and north plenum wall areas (line 20).

The **central system response** is linked to the central system thermal load. The following portions of the system response must be calculated before completing thermal load calculations.

The **fan-coil units** (line 13) were arranged for continuous fan-motor operation during normal system operating hours. (If the motors were cycled by the local room thermostats, it would have been better to proportion the fan motor power in the thermal load estimate, using the ratio of actual zone load (line 15) to fan-coil unit capacity multiplied by fan motor power during operation).

The interior zone VAV fan motor power (line 26) is a variable item and was based on estimated room load as follows:

1) Calculate **interior zone VAV fan CFM.**

$$\text{CFM} = \frac{\text{interior VAV area sensible heat BTUH (line 19)}}{\text{(room temp.} - \text{supply air temp.) (1,09)}}$$

$$= \frac{1{,}668{,}532}{(75\text{-}60)\ (1.09)}$$

$$= 102{,}051 \text{ CFM}$$

2) Calculate **ventilation air to perimeter fan-coil units** (from figure SP-16):

North zone	= 11,500
East zone	= 8,600
South zone	= 11,500
West zone	8,600
1st zone	= 2,300
1st lobby zone	= 800
Fan coil vent air	= 43,300 CFM

3) **Total VAV system fan CFM** = Interior VAV CFM + fan-coil vent CFM

= 102,051 + 43,300

= 145,351 CFM

The VAV fan power input was obtained by reference to the fan performance data, using an assumed 4.6 in. H_2O static pressure. It came to 157 BHP. This value was entered in line 26 of the central system thermal load estimate. Since the fan motor was located in the supply air plenum, the heat of the entire motor input power was charged to the estimate – by dividing BHP motor output by assumed motor efficiency. The ventilation CFM for line 27 and 28 was based on supplying about 15 CFM per person, based

Zone Peak Cooling Requirements and Equipment Capacity

Zone	N	E	S	W	1st	1st Lobby
Fan-Coil Units:						
Number of units	285	228	285	228	8	1
Nominal CFM, ea.	300	600	600	600	1200	1600
Motor watts, ea.	80	170	170	170	466	700
GPM 45° Ch. Wtr. ea.	0.85	1.65	1.65	1.65	5	6
Ventilation CFM	11,500	8600	11,500	8600	2300	800
Peak thermal load*	1113	1658	2431	1948	235	44
Vent. cooling cap.*	−188	−141	−188	−141	−38	−13
Req'd. fan-coil cap.*	925	1517	2243	1807	197	31
Req'd. fan-coil unit*	3.2	6.6	7.9	7.9	24.6	31
Actual fan-coil capacity*	4.2	8.3	8.3	8.3	26	32

*Loads and capacities in MBH sensible heat

Figure SP-16

Peak Central System Cooling Load at 4 pm August

Line	Item	Quantity		GR/# Diff. & Temp. Diff. SHGF		"U" S.C.	Constant		Factors		BTUH
	Perimeter Spaces:										
1	N Solar glass	20,818 sq. ft.	×	11	×	.53				=	121,369
2	E Solar glass	13,566 sq. ft.	×	11	×	.53				=	79,090
3	S Solar glass	17,290 sq. ft.	×	13	×	.53				=	119,128
4	W Solar glass	13,566 sq. ft.	×	165	×	.53				=	1,186,347
5	Glass trans.	65,240 sq. ft.	×	(94-75)	×	1.1				=	1,363,516
6	N Wall trans.	6,308 sq. ft.	×	15°	×	.16				=	15,139
7	E Wall trans.	4,560 sq. ft.	×	17°	×	.16				=	12,403
8	S Wall trans.	6,308 sq. ft.	×	31°	×	.16				=	31,288
9	W Wall trans.	4,560 sq. ft.	×	31°	×	.16				=	22,618
10	1st Floor trans.	7,764 sq. ft.	×	14°	×	.32				=	34,783
11	Lights-to-rm.	429 KW	×				3413	×	75% to rm. × .9 Diversity	=	988,320
12	Equipment	53 KW	×				3413	×	.5 Diversity	=	90,445
13	Fan-coil motors	153 KW	×				3413			=	522,189
14	People, SH.	1,351	×				245	×	.8 Diversity	=	264,796
15									**Fan-Coil Area Sensible Heat**	=	**4,851,431**
	Interior VAV Area:										
16	Lights-to-rm.	543 KW	×				3413	×	75% to Rm. × .9 Diversity	=	1,250,950
17	Equipment	63 KW	×				3413	×	.5 Diversity	=	107,510
18	People, SH.	1,582	×				245	×	.8 Diversity	=	310,072
19									**Interior VAV Area Sensible Heat**	=	**1,668,532**
	Central Apparatus & Return Air Loads:										
20	N Plenum wall trans.	9,042 sq. ft.	×	15°	×	.16				=	21,701
21	E Plenum wall trans.	6,042 sq. ft.	×	17°	×	.16				=	16,434
22	S Plenum wall trans.	7,866 sq. ft.	×	31°	×	.16				=	39,015
23	W Plenum wall trans.	6,042 sq. ft.	×	31°	×	.16				=	29,968
24	21st Floor trans.	13,044 sq. ft.	×	19°	×	.39				=	96,656
25	Lights-to-plenum	972 KW	×				3413	×	25% to Pl. × .9 Diversity	=	746,423
26	VAV fan motor	157 BHP	×				2545	÷	.92 Motor Eff.	=	434,310
27	Vent. air SH	35,000 CFM	×	(94°-75°)	×		1.09			=	724,850
28	Vent. air LH	35,000 CFM	×	(110-65) GR/#	×		.68			=	1,071,000
29	People, LH	2,933	×				205	×	.8 Diversity	=	481,012
									Central Apparatus & Return Air Loads	=	**3,661,369**
30									**Cooling Coil Load**	=	**10,181,332**
31	Chilled wtr. pump	48 KW			×		3413	×	90% Motor Effic.	=	147,442
									Refrigeration Machine Load	=	**10,328,774**
							÷ 12,000			=	**861 tons**

Figure SP-17

on the diversified block load building population of 2933 x 0.8, or 2346 people. This ventilation CFM was held constant during occupied periods, both cooling and heating.

The **chilled water pump motor** operates at a uniform power input whenever the refrigeration machine is on. That is because the fan-coil units and central cooling coils utilize 3-way control valves to keep water flow constant at all load conditions. The pump motor power input (line 31) was estimated using measured amperage to the motor:

$$\frac{\text{Amps} \times \text{volts} \times \text{power factor} \times 1.73}{1{,}000} = \text{KW}$$

$$\frac{69.9 \times 467 \times .85 \times 1.73}{1{,}000} = 48.0 \text{ KW}$$

Unlike the VAV supply fan motor, power loss of the chilled water pump motor windings was dissipated to the equipment room. Only the motor power output was dissipated into the chilled water flow. As a result, in calculating the **central system refrigeration load,** measured power input was reduced by the estimated motor efficiency.

The remaining items of system response and equipment selection did not affect the thermal load estimate. They are needed, however, to **compare central equipment capacity with actual requirements.** Comparisons of estimated requirements and actual capacities are shown in **figure SP-18.** Installed capacity of the **VAV supply fan** was obtained from the original design drawings. Estimated required power input, used previously for the thermal load estimate, was based on operation at the existing fan RPM, with air flow and pressure controlled by the fan inlet guide vanes. It would seem that the fan is operating at an unnecessarily high RPM; if fan speed were reduced to provide wide open inlet vanes at the estimated requirements, power would drop to 124 BHP. This is a worthwhile modification.

The next task is to determine the **central VAV system cooling coil load,** by subtracting the fan-coil unit chilled water load from the total building cooling coil load. Fan-coil unit load is the actual load in the fan-coil areas minus cooling provided by cold ventilation air from the interior VAV system.

Fan-coil area load	=	4,851,431 (line 15)
Ventilation cooling = 43,300 CFM × 1.09 (60°F–75°F)	=	–707,955
Net fan-coil load	=	4,143,476 BTUH
Total cooling coil load	=	10,181,332 (line 30)
minus net fan-coil load	=	–4,143,476
Central cooling coil load	=	6,037,856 BTUH

The central cooling coil selection was checked for this coil load. A chilled water flow of 755 GPM at 45°F was required. Required chilled water pump flow was then calculated:

Fan-coil units:	N zone = 285 units × 0.85 GPM	= 242
	E zone = 228 units × 1.65 GPM	= 376
	S zone = 285 units × 1.65 GPM	= 470
	W zone = 228 units × 1.65 GPM	= 376
	1st zone = 8 units × 5	= 40
	1st lobby zone = 1 unit × 6	= 6
	Total fan-coil flow	= 1,510
	Central cooling coil flow	= 755
	Chilled water pump flow	= 2,265 GPM

Required pump head was estimated by observing the existing system. The head was measured as 85 ft. H_2O.

Central Equipment Peak Cooling Requirements and Capacities

	Estimated Requirements	**Installed Capacity**
VAV supply fan	145,351 CFM @ 4.6″ H_2O 157 BHP	163,000 CFM @ 5.5″ H_2O 176 BHP (200 HP motor)
Chilled water pump	2265 GPM @ 94 ft. H_2O 67 BHP	2150 GPM @ 85 ft. H_2O 58 BHP (60 HP motor)
Water chiller	861 tons @ 641 KW	900 tons @ 684 KW
Condenser water pump	2583 GPM @ 37 ft. H_2O 34 BHP	2700 GPM @ 50 ft. H_2O 43 BHP (50 HP motor)
Cooling tower	861 tons @ 78° W.B. 6 – 15 HP fans	900 tons @ 78° W.B. 6 – 15 HP fans
Toilet exhaust fan	21,120 CFM @ 0.75″ H_2O 5.1 BHP	21,000 CFM @ 1.0″ H_2O 6.9 BHP (7½ HP motor)
Equipment room exhaust fan	30,847 CFM @ 3/8″ H_2O 7 BHP	30,000 CFM @ 3/8″ H_2O 7.9 BHP (10 HP motor)
Garage exhaust fans (Each, 3 total)	14,995 CFM @ 3/8″ H_2O 3.8 BHP	15,000 CFM @ 3/8″ H_2O 3.8 BHP (5 HP motor)

Note that equipment for this building was carefully sized to match peak load.

Figure SP-18

Pressure drop through the water chiller was measured as 24 ft. H_2O. From equipment ratings this indicated a flow of 2,150 GPM. If flow were increased to the required 2,265 GPM, the required pump head would be:

$$\text{Required head} = \frac{\text{required GPM}^2}{\text{actual GPM}^2} \times \text{actual head} =$$

$$\frac{2{,}265^2}{2{,}150^2} \times 85 = 94 \text{ ft. } H_2O$$

Pump flow could be increased by replacing the pump impeller and the resulting power requirement from the pump would be 56 KW. Since the system seems to be working well, the calculations were clearly conservative and this change is unnecessary.

Performance of the **central refrigeration machine** was checked at the 861 ton requirement, and power input was 641 KW.

Using the 3 GPM per ton typical of this type of machine, the required **condenser water pump flow** is 2,583 GPM. Actual flow estimated from the condenser pressure drop measurement is 2,900 GPM, and power input is 36 KW. Pump flow could be cut to 2,583 GPM by trimming the pump impeller, which would result in a power reduction to 28 KW. This possible reduction of 8 KW would be offset by increased power input to the refrigeration machine, which may be greater or less than 8 KW. To determine the true benefit of this type of change would require a complete refrigeration system analysis (compressor, condenser water pump, and cooling tower performance).

The 900 ton capacity of the **cooling tower** is obviously adequate and imposes no energy penalty, since the cooling tower fans will be automatically controlled to operate only as needed to meet loads.

Toilet exhaust fan requirements were checked, based on a criterion of 2 CFM per sq. ft. of toilet floor area.

$$10{,}560 \text{ sq. ft. area} \times 2 \text{ CFM/sq. ft.} = 21{,}120 \text{ CFM}$$

Using an estimated 0.75 in. static pressure, fan ratings were reviewed to obtain a BHP requirement of 5.1 BHP. The approximate present load on this fan can be estimated by means of a ratio of the actual and nominal voltage and amperage.

$$\frac{\text{Actual BHP}}{\text{Nominal HP}} = \frac{\text{Actual amps}}{\text{Nominal amps}} \times \frac{\text{Actual volts}}{\text{Nominal volts}}$$

$$\frac{\text{Actual BHP}}{7.5} = \frac{9.3}{10.3} \times \frac{467}{460} = 6.9 \text{ BHP}$$

If the estimate of 0.75" static pressure is correct, the fan is probably delivering too much air and the drive should be changed to reduce air flow and energy consumption. This excess air flow has a bad impact on HVAC system energy consumption because it leads to higher ventilation air flow or higher infiltration. Either of these increases energy consumption, both during cooling and heating seasons.

Peak demand KW was estimated from the actual operating condition:

$$\text{KW} = \frac{9.3 \text{ amps} \times 467 \text{ volts} \times 0.8 \text{ PF} \times 1.73}{1{,}000} = 6.0 \text{ KW}$$

The fan operates continuously at the same load, so average energy consumption is the same as peak demand. This 6 KW is used for both Parts B and C of the energy estimate.

Equipment room exhaust fan requirements were estimated using a design temperature of 100°F in the equipment rooms. Major cooling loads are heat gain through the roof and heat from electric motors.

The heat flow through the roof was estimated using roof temperature of 135°F.

$$\text{Load} = 15{,}120 \text{ sq. ft.} \times .15 \text{ "U"} \times (135°\text{F} - 100°\text{F}) = 79{,}380 \text{ BTUH}$$

Note that design temperature assumptions can always be questioned and the conditions creating them can often be changed.

The refrigeration machine is a hermetic centrifugal type in which heat due to motor inefficiency is rejected to the condenser water, not to the equipment room. The boiler and hot water pump are turned off, and the exhaust fans are located on the roof. VAV fan motor heat was absorbed by the interior zone air stream, so the major items of motor heat are:

Chilled water pump	= 48 KW
Condenser water pump	= 36 KW
Elevators	= 176 KW
Total	= 260 KW

The heat rejection of these motors is estimated using an assumed motor efficiency of 90%. Wasted heat is therefore 10% of the total input:

$$10\% \times 260 \text{ KW} \times 3{,}413 \text{ BTUH/KW} = 88{,}738 \text{ BTUH}$$

The total equipment room heat gain is:

Roof load	= 79,380
Motor heat	= 88,738
Total	= 168,118 BTUH

The required ventilation air at 95°F is:

$$\text{CFM} = \frac{168{,}118 \text{ BTUH}}{1.09\ (100°\text{F room} - 95°\text{F vent. air})} = 30{,}847 \text{ CFM}$$

The required motor input from the fan ratings, using an estimated 0.375 in. H_2O fan static pressure, is 7 BHP.

The approximate actual fan BHP is:

$$\text{BHP} = 10 \text{ HP motor} = \frac{10.7 \text{ actual amps}}{13.8 \text{ nominal amps}} \times \frac{467 \text{ actual volts}}{460 \text{ nominal volts}} = 7.9 \text{ BHP}$$

This agrees reasonably well with the estimate, so that equipment room temperatures should be measured during hot summer weather before attempting to reduce the ventilation air quantity.

Peak demand KW of the equipment room exhaust fan was calculated from measured operating conditions and was used in Part B of this estimate.

$$\text{KW} = \frac{10.7 \text{ amps} \times 467 \text{ volts} \times 0.8 \text{ PF} \times 1.73}{1{,}000} = 6.9 \text{ KW}$$

The average energy consumption estimate assumed the equipment room exhaust fan to be operating in hot weather, but controlled by a thermostat so that operation was not required during cooler weather (72°F and below).

Garage ventilation

Requirements for garage ventilation were estimated on the basis of providing 4 air changes per hour.

Garage volume = 74,976 sq. ft. × 9 ft. ceiling = 674,784 cu. ft.

$$674{,}784 \text{ cu. ft.} \times \frac{4 \text{ air changes}}{60 \text{ min.}} = 44{,}986 \text{ CFM}$$

This compares favorably with the three 15,000 CFM fans as installed.

Operation of these fans is not related to outside temperature or to the operating schedule of the HVAC system. Their energy consumption was therefore estimated using operating rate times hours of operation, rather than including them in the HVAC system bin method estimate.

Actual power input to each of the three garage exhaust fans was measured as 5.2 amps when the supply voltage was 467 volts. These fans operate on the same schedule as the garage lights, namely, 15 hours per day, Monday through Friday. Peak demand is the same as average hourly consumption, that is, 10 KW.

Power input and monthly usage calculations for garage fans, shown in **figure SP-19,** were posted to the monthly HVAC peak demand summary in **figure SP-28** and to the monthly energy consumption summary in **figure SP-41.**

Garage Exhaust Fans

Peak Demand:

$$\text{KW Input} = 3 \text{ Fans} \times \frac{5.2 \text{ amps} \times 467 \text{ volts} \times .8 \text{ PF}}{1000} \times 1.73 = 10.1 \text{ KW}$$

Daily Consumption:

10.1 KW × 15 hrs/day = 151 KWH per occupied day

Monthly Consumption:

Feb. = 151 × 5/7 days/wk × 51/52 wks/yr × 28 days = 2962 KWH
30 Day Months = 151 × 5/7 days/wk × 51/52 wks/yr × 30 days = 3173 KWH
31 Day Months = 151 × 5/7 days/wk × 51/52 wks/yr × 31 days = 3279 KWH

Note that if the garage ventilation fan operates to reduce carbon monoxide buildup, it may only need to do so a few hours each day; a sensor can sense the buildup.

Figure SP-19

Step A3 Zone heating requirements

This step is designed to verify zone peak heating capacities. Time of peak load was assumed to be at 8 AM in January, before a particular office was occupied, but after the central ventilation system was delivering cold outside ventilation air. Infiltration of outside air through window frames was also included. Zone thermal load and zone system response for the east zone are both shown in **figure SP-20.**

East Zone Peak Heating Estimate 8 AM January at 10° Outside

Line	Item	Quantity		Temp. Diff. SHGF		"U" S.C.	Constant	Factors		BTUH
1	Glass solar	13566 sq. ft.	×	0						0
2	Glass trans.	13566 sq. ft.	×	(10-75)	×	1.1			=	– 969,969
3	Wall trans.	4560 sq. ft.	×	(10-75)	×	.16			=	– 47,424
4	Plenum wall trans.	6042 sq. ft.	×	(10-75)	×	.16			=	– 62,837
5	Infiltration	3200 CFM	×	(10-75)	×		1.09		=	– 226,720
								Zone Peak Heat Loss	=	**– 1,306,950**
6	Zone vent. cooling	8600 CFM	×	(60-75)			1.09		=	– 140,610
7	Fan-Coil motors	38.8 KW					3413		+	132,400
							Required fan-coil capacity		=	– 1,315,160

÷ 228 units = 5768 BTUH per unit
actual capacity = 10,500 BTUH
with 0.5 GPM of 160° water

Figure SP-20

The glass solar gain (line 1) was assumed to be zero due to the early morning winter hour and low sun, if any.

Infiltration (line 5) was assumed to be one air change per hour, and the cooling effect of the cold ventilation air was based on the design flow rate of 0.4 CFM per sq.ft. of floor area.

The results of all zone peak heating estimates and their comparison with existing fan coil unit capacities are shown in **figure SP-21.** Much excess capacity is available for the east, south and west units. The system hot water temperature is determined by the north and first floor units, both of which require 160°F hot water.

Step A4 Central system heating requirements

This verifies central system peak heating loads and capacities. Assumptions for this estimate are: 1) The interior zone load is zero. Therefore, the central VAV supply fan CFM is the exterior zone ventilation CFM plus an estimated 10% VAV terminal unit leakage. 2) The hot water pump is operating. 3) The chilled water pump and refrigeration system are off.

The central VAV fan CFM is:

Exterior zone ventilation	= 43,300
VAV terminal unit leakage (10% of 102,051)	= 10,205
Total	= 53,505 CFM

From the fan ratings, power input is 58 BHP. Assuming motor efficiency drops to 80%, the power input and fan motor heat is:

Exterior Zone Design Heating Capacities

Zone	N	E	S	W	1st	1st Lobby
Zone peak heat loss MBH	1,688	1,307	1,688	1,307	389	140
Central vent. CFM	11,500	8,600	11,500	8,600	2,300	800
Vent. cooling MBH	188	141	188	141	38	13
Fan KW	22.8	38.8	48.5	38.8	3.7	0.7
Fan heat credit MBH	77.8	132.3	165.4	132.3	12.6	2.4
Fan-coil load MBH	1,798	1,316	1,711	1,316	414	151
No. of units	285	228	285	228	8	1
Load per unit MBH	6.3	5.8	6.0	5.8	51.8	151
Actual capacity MBH	7.2	10.4	10.4	10.4	53.4	151
GPM 160° water	0.5	0.5	0.5	0.5	8	15

Figure SP-21

$$\frac{58 \text{ BHP} \times .745 \text{ KW/HP}}{0.80 \text{ motor efficiency}} = 54.1 \text{ KW} \times 3413 = 184{,}600 \text{ BTUH}$$

Temperature rise of the supply air due to fan motor heat is:

$$\frac{184{,}300 \text{ BTUH}}{(1.09)\,(53{,}505 \text{ CFM})} = 3.2°\text{F}$$

Supply air consists of 35,000 CFM of outside air at 10°F and 18,505 CFM of return air at 75°F. Temperature of the mixture is 32.5°F. It must be heated to 56.8°F which, including the 3.2°F fan motor heat, will provide 60°F supply air leaving the fan. Heating coil input is:

53,505 CFM (1.09) (56.8 – 32.5)
= 1,417,187 BTUH ÷ 3,413
= 415 KW

A check of the existing coil showed that a hot water flow of 70 GPM at a160°F would provide the required capacity. **The hot water pump flow requirement is the central heating coil water flow** of 70 GPM plus the total fan-coil unit GPM of 536 (GPM per unit is shown in figure SP-21), or 606 GPM. The present motor load is 12 BHP, with 10 KW power input. The useful work of this power will be converted to useful heat, and will be a credit to the boiler input. Thus

10KW × 90% motor efficiency × 3,413 = 30,700 BTUH

The total fan-coil unit heating load was 6,706 MBH. Deducting the 31 MBH hot water pump motor credit, and converting to KW input, gives the boiler load:

Fan-coil total load	= 6,706
– pump credit	= 31
Net load	= 6,675 MBH ÷ 3.41
	= 1,957 KW

Thus **total boiler load** is 1,957 KW (fan-coil) + 415 KW (interior VAV system), or 2,372 KW. There is no need to add insulation losses from hot water piping because most of this is located within the building and any heat losses would tend to offset some of the building's heat load.

A comparison of theoretical loads and actual equipment capacities is shown in **figure SP-22.**

Actual boiler KW is only 2,000 KW. The boiler was probably undersized expressly to cut back on electric demand, on the theory that one would really not need to provide 35,000 CFM of outside air for ventilation this early in the morning.

The excess **hot water pump capacity** has little effect on the HVAC system's energy consumption because most of the additional power input is a direct credit against the boiler load.

Central Equipment Peak Heating Requirements and Capacities

	Estimated Requirement	**Installed Capacity**
Electric boiler	2372 KW	2000 KW
Hot water pump	606 GPM @ 40 ft. H_2O 8 BHP	680 GPM @ 50 ft. H_2O 12 BHP (15 HP motor)

Figure SP-22

Peak Central System Cooling Demand Load at 3 PM March, 77 D.B., 61 W.B.

Line	Item	Quantity		Temp. Diff. SHGF		"U" S.C.	Constant		Factors		BTUH
	Perimeter Spaces:										
1	N Solar glass	20,818 sq. ft.	×	16	×	.53				=	176,537
2	E Solar glass	13,566 sq. ft.	×	16	×	.53				=	115,040
3	S Solar glass	17,290 sq. ft.	×	79	×	.53				=	723,932
4	W Solar glass	13,566 sq. ft.	×	189	×	.53				=	1,358,906
5	Glass trans.	65,240 sq. ft.	×	(77-75)	×	1.1				=	143,528
6	N Wall trans.	6,308 sq. ft.	×	−2	×	.16				=	− 2,019
7	E Wall trans.	4,560 sq. ft.	×	0	×	.16				=	0
8	S Wall trans.	6,308 sq. ft.	×	14	×	.16				=	14,130
9	W Wall trans.	4,560 sq. ft.	×	14	×	.16				=	10,214
10	1st Floor trans.	7,764 sq. ft.	×	−3	×	.32				=	− 7,453
11	Lights-to-rm.	429 KW	×				3413	×	75% to rm. × .9 Diversity	=	988,320
12	Equipment	53 KW	×				3413	×	.5 Diversity	=	90,445
13	Fan-coil motors	153 KW	×				3413			=	522,189
14	People, SH.	1,351	×				245	×	.8 Diversity	=	264,796
15									**Fan-Coil Area Sensible Heat**	=	**4,398,565**
	Interior VAV Area:										
16	Lights-to-rm.	543 KW	×				3413	×	75% to Rm. × .9 Diversity	=	1,250,950
17	Equipment	63 KW	×				3413	×	.5 Diversity	=	107,510
18	People, SH.	1,582	×				245	×	.8 Diversity	=	310,072
19									**Interior VAV Area Sensible Heat**	=	**1,668,532**
	Central Apparatus & Return Air Loads:										
20	N Plenum wall trans.	9,042 sq. ft.	×	−2	×	.16				=	− 2,893
21	E Plenum wall trans.	6,042 sq. ft.	×	0	×	.16				=	0
22	S Plenum wall trans.	7,866 sq. ft.	×	14	×	.16				=	17,620
23	W Plenum wall trans.	6,042 sq. ft.	×	14	×	.16				=	13,534
24	21st Floor trans.	13,044 sq. ft.	×	2°	×	.39				=	10,174
25	Lights-to-plenum	972 KW	×				3413	×	25% to Pl. × .9 Diversity	=	746,423
26	VAV fan motor	157 BHP	×				2545	÷	.92 Motor Eff.	=	434,310
27	Vent. air SH	35,000 CFM	×	(77-75)		×	1.09			=	76,300
28	Vent. air LH	35,000 CFM	×	(55-65) GR/#		×	.68			=	− 238,000
29	People, LH	2,933	×				205	×	.8 Diversity	=	481,012
									Central Apparatus & Return Air Loads	=	**1,538,480**
30									**Cooling Coil Load**	=	**7,605,577**
31	Chilled wtr. pump	48 KW			×		3413	×	90% Motor Effic.	=	147,442
									Refrigeration Machine Load	=	**7,753,019**
							÷ 12,000			=	**646 tons**

Figure SP-23

Part B Calculate HVAC peak energy demand

The monthly electric utility cost includes a demand charge, so a peak demand profile is needed to obtain monthly demand values.

Step B1 Calculate and plot profile points – cooling demand

Calculating and plotting the cooling demand profile calls for computing an additional profile point to that calculated when verifying peak cooling requirement in Steps A1 or A2. Because the building uses electricity for heating, the demand load in the winter months will almost certainly occur at the peak heating rather than cooling situation.

Since peak cooling condition has already been computed, March is chosen as an intermediate profile point, as it is a typical spring month in which the cooling load might set the monthly demand. The block load format could be used because outside temperature is above the temperature at which heating is required in the exterior zones.

The calculations are done in the same manner as in Step A2 **(figure SP-17),** except the outside weather conditions and solar values are for March. The highest temperature bin in March with a reasonable number of hours is 77°F dry bulb, with a 61°F mean coincident wet bulb temperature. **Figure SP-23** shows the 77°F profile point peak demand calculation.

The central VAV fan and chilled water pump HP are the same as for the 94°F point calculated in Step A2. **Refrigeration system power consumption** is obtained from the equipment ratings.

Figure SP-24 shows the tabulation of power inputs at both the 94°F and 77°F profile points, as well as plotting of the profile.

Peak Cooling Demand Profile

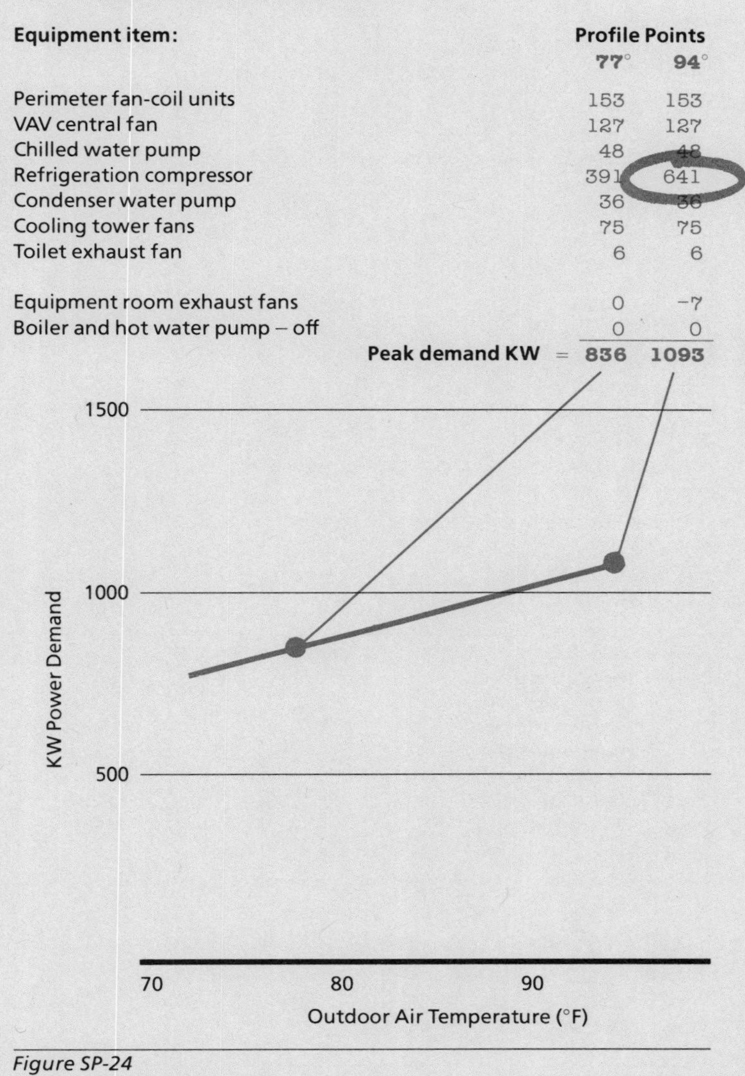

Equipment item:	Profile Points 77°	94°
Perimeter fan-coil units	153	153
VAV central fan	127	127
Chilled water pump	48	48
Refrigeration compressor	391	641
Condenser water pump	36	36
Cooling tower fans	75	75
Toilet exhaust fan	6	6
Equipment room exhaust fans	0	−7
Boiler and hot water pump – off	0	0
Peak demand KW =	**836**	**1093**

Figure SP-24

Central VAV System Profile Point Determination

Temperature rise of supply air because of fan motor heat:

$$= \frac{184{,}600 \text{ BTUH}}{1.09\,(55{,}505 \text{ CFM})} = 3.2^\circ$$

Mixed air temperature entering fan $= 60^\circ - 3.2^\circ = 56.8^\circ$

Let x = outside air temperature at which mixed air temperature will equal 56.8°

$$\frac{18{,}505 \text{ return air CFM}}{53{,}505 \text{ supply air CFM}} \times (75^\circ - x) + x = 56.8^\circ$$

$\mathbf{x = 47.2^\circ}$

Figure SP-25

Step B2 Calculate and plot profile points — heating demand

A heating demand profile must be prepared because the electric boiler will establish the peak demand during winter months.

The two major components of the boiler load are the fan-coil unit heating load and the central VAV heating coil load.

Figure SP-20 shows that all fan-coil zone heat loss elements are related to temperature difference between room and outside temperatures. Therefore, the load will be zero when the outdoor air temperature equals the 75°F room temperature. The load loss is reduced by the fan-coil unit and hot water pump motor heat. Reviewing the central VAV heating coil loads in Step A4 shows that this load consists of heating the outside and return air mixture to 60°F. This portion of the boiler load is, therefore, zero when mixed air temperature plus fan heat equal 60°F.

Two additional profile points are thus needed to plot the heating demand profile: one at about 75°F, the other at about 60°F when the central mixed air no longer needs heating.

How to calculate the outside temperature at which central VAV heating is no longer needed is shown in **figure SP-25.** It comes to 47.2°F. This temperature was selected as the third profile point. The fan-coil unit boiler demand calculations are shown in **figure SP-26.**

The summation of equipment power input at the heating demand profile points and the completed profile are shown in **figure SP-27.** The 75°F point is theoretical and is used only to establish the slope of the line. That is why the chilled water pump and refrigeration system are "OFF," when they would normally be "ON" at 75°F. The building operating schedule calls for the chilled water pump to be started after the building is occupied and after the heat from lights and sun begins to generate a cooling load on the fan-coil units.

Fan-Coil Area Boiler Load at Profile Points

	Profile Points	
	47.2°	75°
Exterior zone heat loss	− 2788*	0
Central ventilating system cooling	− 708	− 708
Fan-coil unit motor heat	+ 522	+ 522
Hot water pump motor heat	+ 31	+ 31
Boiler Load	= 2943 MBH	155 MBH
÷ 3.41	= 863 KW	45 KW

*Heat loss at 10° Outside = 6519 MBH
(Sum of Zone Peak Heat Losses from Fig. SP-21)

$$\text{Loss at } 47.2° = 6519\left(\frac{75-47.2}{75-10}\right) = 2788 \text{ MBH}$$

Figure SP-26

Peak Heating Demand Profile

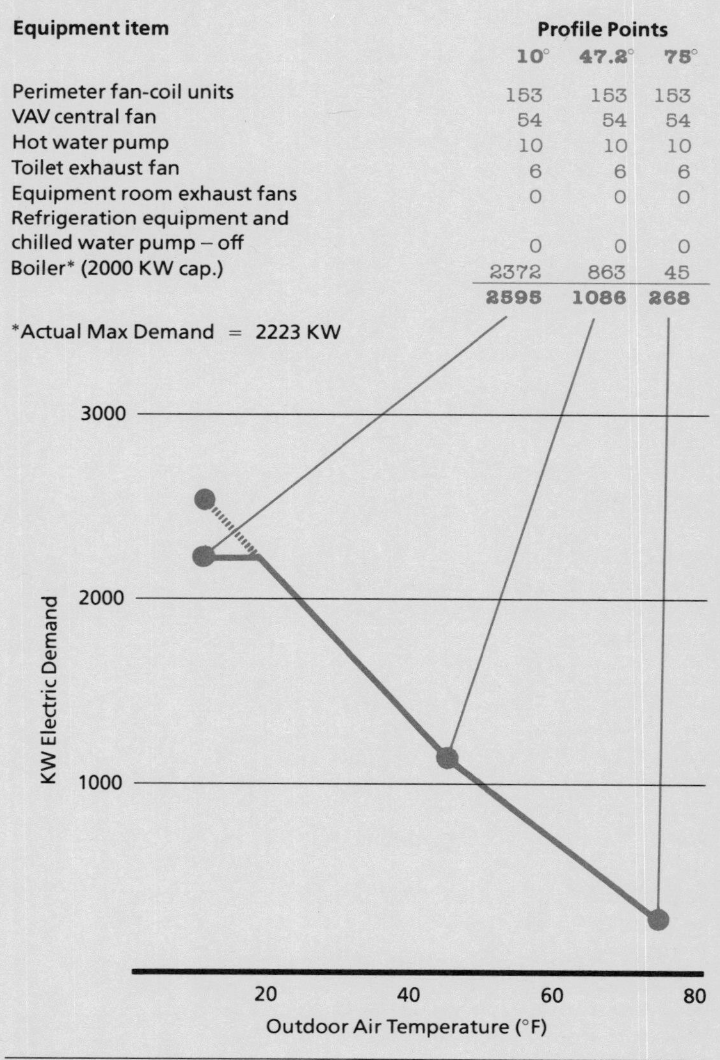

Equipment item	Profile Points		
	10°	47.2°	75°
Perimeter fan-coil units	153	153	153
VAV central fan	54	54	54
Hot water pump	10	10	10
Toilet exhaust fan	6	6	6
Equipment room exhaust fans	0	0	0
Refrigeration equipment and chilled water pump – off	0	0	0
Boiler* (2000 KW cap.)	2372	863	45
	2595	1086	268

*Actual Max Demand = 2223 KW

Figure SP-27

Step B3 Monthly peak demand summary

The monthly peak demand summary is shown in **figure SP-28.** It is obtained by selecting the highest and lowest temperatures from the monthly bin weather data, and entering the peak demand profile. The highest temperature is used with the cooling profile, and the lowest temperature with the heating.

The highest of the cooling or heating demand values for each month is posted to the monthly total building summary for calculating energy cost.

Part C Calculate HVAC average energy consumption

When calculating energy consumption, one must break down operation of the system into occupied and unoccupied periods. Separate energy profiles and energy consumption summaries must be prepared for each period.

Step C1 Calculate and plot profile points/ average energy consumption during occupied period

This step consists of preparing the average energy consumption profile for the occupied period. It is clear from examining the temperature bin data that most of the occupied period operating hours in the Atlanta, Ga., area occur between 30°F and 90°F. Accordingly, 32°F (from the 30-34°F bin) and 92°F (from the 90-95°F bin) were selected as profile points.

The central VAV system has an outside air economizer cycle which now operates with minimum ventilation air down to 70°F, when the enthalpy-type outside air economizer cycle switches to 100% outside air. A profile point at 72°F was chosen to identify any change in profile slope due to enthalpy control. Outside air ceases to be a refrigeration load at about 50°F, when return air damper leakage and fan motor heat result in about 60°F supply air without refrigeration. A 52°F profile point was chosen for this reason.

Calculating the profile point for energy consumption in occupied periods calls for two types of estimates. Heating energy will usually not be required either at the 92°F or 72°F points, so the block load method will be used. Both heating and cooling may be required at the 52°F and 32°F points, so the hour-by-hour, zone-by-zone method will be used at these points.

Figure SP-29 was prepared from the bin weather data to determine the month that most nearly represents each profile point. The per cent sunshine and mean coincident wet bulb temperature for the selected month were used for computing load at that profile point.

The block load format implies that all loads must be reduced to an ***average*** *load that would exist over the* ***entire operating day.*** *The load must be completely independent of time of day. To do this, solar loads for each exposure for the entire day must be averaged, as shown in* ***figure SP-30.***

Monthly Peak Demand Summary

Month	Temp Peak	Profile KW*	Garage fan KW	Total KW
January	7° (H)	2223	10	2233
February	7° (H)	2223	10	2233
March	12° (H)	2223	10	2233
April	32° (H)	1783	10	1793
May	42° (H)	1377	10	1387
June**	97° (C)	1093	10	1103
July**	97° (C)	1093	10	1103
August**	97° (C)	1093	10	1103
September	47° (H)	1174	10	1184
October	27° (H)	1986	10	1996
November	17° (H)	2223	10	2233
December	7° (H)	2223	10	2233

*Includes all HVAC equipment except garage exhaust fans
**The peak design condition of 94° was exceeded.

Figure SP-28

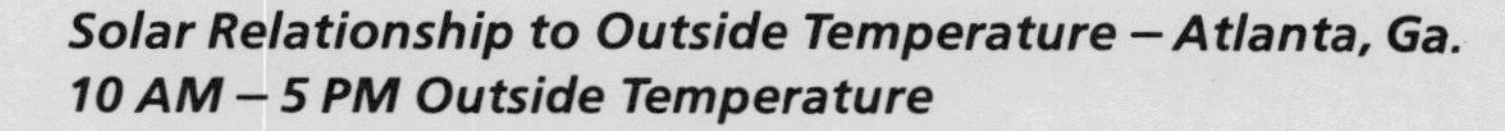

Solar Relationship to Outside Temperature – Atlanta, Ga. 10 AM – 5 PM Outside Temperature

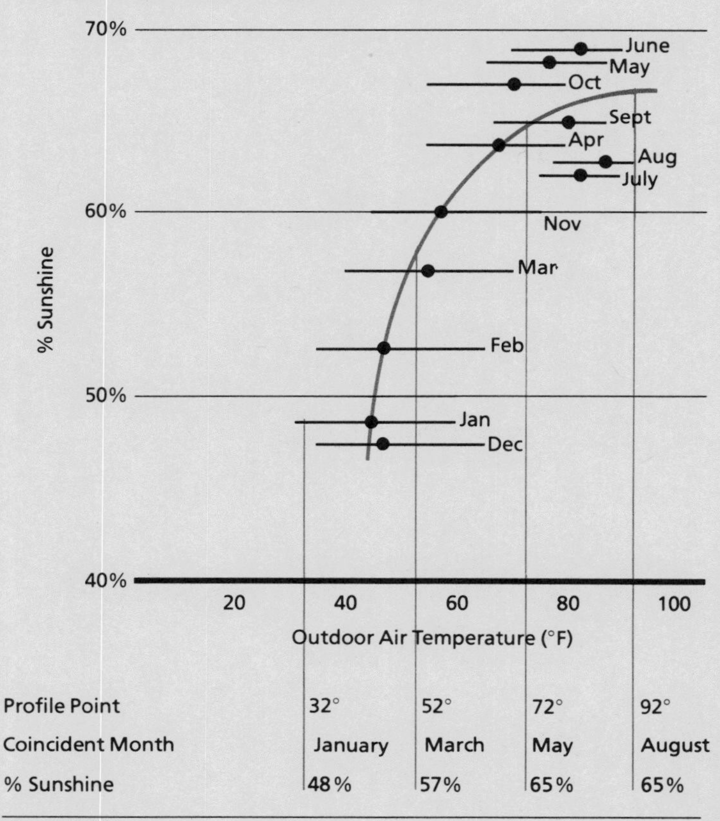

Figure SP-29

The block load estimate for the 72°F profile point is shown in **figure SP-31.** Note that in a consumption analysis wall transmission for all exposures is lumped together and based on actual, not on equivalent temperature difference. Outside temperature will be rising during some hours and falling in others, which tends to equalize any difference.

Heat transmission through the 21st floor (line 18) was based on an equipment room temperature 5°F above the outside temperature.

The VAV fan motor power was estimated in the same manner as for the peak cooling load estimate in figure SP-17. Thus

$$\text{VAV area SH (line 16)} = 838{,}126$$

$$\text{VAV area CFM} = \frac{838{,}126}{(1.09)\,(15^\circ\text{F temp. dif.})} = 51{,}262$$

$$\text{Fan-coil ventilation} = 43{,}300$$

$$\text{Total VAV fan} = 94{,}562 \text{ CFM}$$

The 94 BHP fan power for line 20 was obtained from the manufacturer's fan performance ratings using this air quantity.

The 92°F profile point was calculated in the same way, and the refrigeration machine load was 544 tons.

Average Daily Solar Loads For May

		Solar exposure			
Hour		N	E	S	W
5		22	30	2	2
6		22	100	4	4
7		20	155	9	9
8		14	164	12	12
9		13	145	14	13
10		14	99	20	14
11		14	44	27	14
12		14	14	30	14
1		14	14	27	44
2		14	14	20	99
3		13	13	14	145
4		14	12	12	164
5		20	9	9	155
6		22	4	4	100
7		22	2	2	30
Total		252	819	206	819
16 Hour avg.	=	16	51	13	51
Corrected avg.*	=	18	56	14	56

*Correction factor = .9 haze × 1.049 dew point × 1.17 sash area = 1.10

Figure SP-30

Average Block Load May – 72° DB, 64° WB

Line	Item	Quantity	Temp. Diff. SHGF	"U" S.C.	Constant	Factors	BTUH
	Fan Coil Area:						
1	N Solar Glass	20,818 sq. ft. ×	18 ×	.53		× 65% Sun	= 129,092
2	E Solar Glass	13,566 sq. ft. ×	56 ×	.53		× 65% Sun	= 261,715
3	S Solar Glass	17,290 sq. ft. ×	14 ×	.53		× 65% Sun	= 83,390
4	W Solar Glass	13,566 sq. ft. ×	56 ×	.53		× 65% Sun	= 261,715
5	Glass Trans.	65,240 sq. ft. ×	(72-75) ×	1.1			= – 215,292
6	Wall Trans.	21,736 sq. ft. ×	(72-75) ×	.16			= 10,433
7	1st Floor Trans.	7,764 sq. ft. ×	(72-75) ×	.32			= – 7,453
8	Lights-to-Rm.	429 KW ×			3413	× 75% to Rm. × .8 Div. × 10/16 hrs	= 549,066
9	Equipment	53 KW ×			3413	× .25 Div. × 8/16 hrs	= 22,611
10	People, SH.	1,351 ×			245	× .6 Div. × 8/16 hrs	= 99,299
11	Fan-Coil Motors	153 KW ×			3413		= 522,189
12						**Fan-Coil Area Sensible Heat**	= **1,695,899**
	Interior VAV Area:						
13	Lights-to-Rm.	543 KW ×			3413	× 75% to Rm. × .8 Div. × 10/16 hrs	= 694,972
14	Equipment	63 KW ×			3413	× .25 Div. × 8/16 hrs	= 26,877
15	People, SH.	1,582 ×			245	× .6 Div. × 8/16 hrs	= 116,277
16						**VAV Area Sensible Heat**	= **838,126**
	Central Apparatus & Return Air Loads:						
17	Plenum Wall Trans.	28,992 sq. ft. ×	(72-75) ×	.16			= – 13,916
18	21st Floor Trans.	13,044 sq. ft. ×	(77-75) ×	.39			= 10,174
19	Lights-to-Plenum	972 KW ×			3413	× 25% to Pl. × .8 Div. × 10/16 hrs	= 414,680
20	VAV Fan Motor	94 BHP ×			2545	÷ .90 Motor Eff.	= 265,811
21	Vent. Air, SH	35,000 CFM ×	(72-75) ×		1.09		= – 114,450
22	Vent. Air, LH	35,000 CFM ×	(76-65) GR/# ×		.68		= 261,800
23	People, LH	2,933 ×			205	× .6 Div. × 8/16 hrs	= 180,380
24						**Central Apparatus & Return Air Loads**	= **1,004,479**
25						Cooling Coil Load	= 3,538,504
26	Chilled Wtr. Pump	48 KW		×	3413	× 90% Motor Effic.	= 147,442
						Refrigeration Machine Load	= **3,685,946**
						÷ 12,000	= **307 tons**

Figure SP-31

The hour-by-hour format for the 52°F and 32°F profile points requires a separate estimate for each zone. **Figure SP-32** shows the east zone estimate for the 52°F point.

The operating schedule for lights called for 10 hours' operation per day. This was done using the hour-by-hour format by showing 9 hours of normal operation plus 5 hours of operation at 20% of the normal rate, to account for evening cleaning crews and evening overtime work.

Note that this zone required 1,744 MBH of cooling energy and 2908 MBH of heating energy during the 16-hour HVAC system operating period. If the block load format were used, the 1,744 MBH of cooling load would have cancelled 1,744 MBH of heating load, leaving only 1,164 MBH of heating energy required, thereby underestimating the heating load by 60% and the cooling energy by 100%.

Hour-by-Hour Zone Summary

	52° Profile Point		32° Profile Point	
Zone	**Cool**	**Heat**	**Cool**	**Heat**
North	0	396	0	859
East	109	182	0	500
South	126	228	57	448
West	111	184	0	500
Exterior-1	0	95	0	207
Lobby-1	0	36	0	68
Total zone loads	346	1121	57	2582
Chilled water pump	148		148	
Hot water pump		−31		−31
Equipment loads	494	1090	205	2551
	41 Tons	**320 KW**	**17 Tons**	**748 KW**

Note: All Loads in MBH

Figure SP-33

Typical Hour-by-Hour Zone Estimate

Profile Point = 52° Zone = East
Month = March % Sun = 57

Wall Trans. = 4560 Sq. Ft. × .16 "U" × (52°-75°) = − 16,781 BTUH
Glass Trans. = 13566 Sq. Ft. × 1.1 "U" × (52°-75°) = − 343,220 BTUH
Infiltration = 1600 CFM × 1.09 (52°-75°) = − 40,112 BTUH
Transmission + Infiltration = − 400,113 BTUH
= − 400 MBH

Max. Lights = 86 KW × 3.413 × .75 pl × .8 Diversity = 176.1 MBH
Max. Equip. = 11 KW × 3.413 × 0.25 Diversity = 9.5 MBH
Max. People = 268 × 0.245 × 0.6 Diversity = 39.4 MBH
Fan-Coil Motors = 38.8 KW × 3.413 = 132.4 MBH
Solar = 13,566 sq. ft. × 57% Sun × .53 S.C. × .9 Haze ÷ 1000 = 3.688 × Corrected SHGF for Time of Day

	Zone Loads						Fan-Coil Energy			
							Fan-Coil	Central		
Hour	T + I	Lights	Equip.	People	Solar	Total	Motors	Vent.	Cool	Heat
6	−400	0	0	0	0	−400	132	−141	0	409
7		0	0	0	621	221			212	0
8		176	10	39	791	616			607	
9					721	546			537	
10					516	341			332	
11					240	65			56	
12			0	0	70	−154			0	163
1			10	39	70	−105				114
2					65	−110				119
3					60	−115				124
4					50	−125				134
5		35	0	0	25	−340				349
6					0	−365				374
7										
8										
9										

Day Total = 1744 2908
÷ 16 hours = Hourly Average Load = 109 182

Note: All Loads and Capacities in MBH.

Figure SP-32

Sample Problem

The summary of individual hour-by-hour zone estimates for the 52°F and 32°F profile points is shown in **figure SP-33.** (Actual individual estimates are not shown.)

The interior zone loads and fan power are the same for all profile points. There is no **refrigeration load** at the 52°F and 32°F profile points due to the outside air economizer cycle. The need for **heating** was checked by calculating outside and return air mixture temperature at the 32°F profile point, as shown in **figure SP-34.**

The mixed air temperature was above the 60°F supply air temperature required, so the economizer control will have to open to admit more than the 35,000 CFM of ventilation air. At this average condition, preheat will not be required. Note that if an hour-by-hour calculation were made, the load during early morning and late evening hours would not have been enough to avoid preheat; by this averaging process a slight error has, therefore, been introduced.

Electric power input to the refrigeration machine cooling tower fan was obtained by referring to the equipment ratings.

Energy consumption of the equipment is totaled for each profile point, and the points are plotted in **figure SP-35.**

Occupied Period Average Energy Profile

Equipment Item	Profile Points 32°	52°	72°	92°
Perimeter fan-coil units	153	153	153	153
VAV central fan	78	78	78	78
Chilled water pump	48	48	48	48
Refrigeration compressor	40	88	218	353
Condenser water pump	11	25	36	36
Cooling tower fan	0	0	54	75
Hot water pump	10	10	0	0
Boiler	748	320	0	0
Toilet exhaust fan	6	6	6	6
Equipment room exhaust fans	0	0	0	7
Total KW =	**1094**	**728**	**593**	**756**

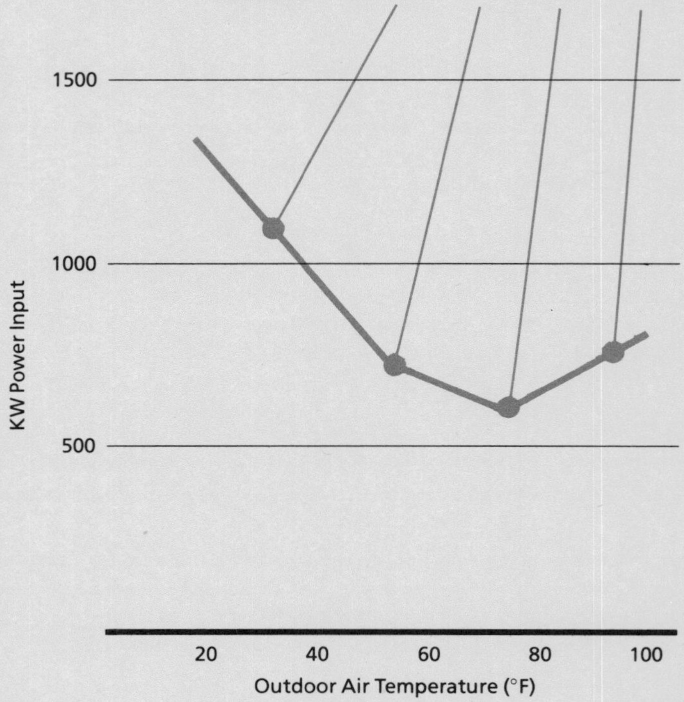

Figure SP-35

Mixed Air Temperature at 32° Profile Point

Central system return air loads:

Plenum wall transmission: 28,992 sq. ft. × (32° − 75°) × .16 "U" = − 199,465

21st floor transmission: 13,044 sq. ft. × (32° − 52°) × .39 "U" = − 101,743

Lights-to-plenum: 972 KW × 3413 × 25% × .8 Div. × $10/16$ hrs = + 414,680

Total = **113,472**

Return air temperature gain = 113,472 ÷ (1.09 × 59,562 CFM) = 1.8°

Return air temperature = 75° + 1.8° = 76.8°

Mixture temperature of 35,000 CFM outside air at 32° with 59,562 CFM return air at 75.8°:

$$\frac{\text{59,562 CFM return air}}{\text{94,562 CFM supply air}} \times (76.8° - 32°) = 28.2°$$

28.2° + 32° = 60.2°

Figure SP-34

Step C2 Calculate and plot profile points/ average energy consumption during unoccupied period

Preparing an energy profile for the unoccupied period is like doing it for the occupied period. Building transmission and infiltration losses during the unoccupied period will be zero at the room temperature (75°F), so this was chosen as one point. Night temperatures are colder than daytime, so 12°F was chosen as an extreme average heating point. The equipment room does not require heating above 52°F outside temperature; it was therefore selected as a third point.

Cooling is not available during the unoccupied period, so a simple block load estimate format is used. **Figure SP-36** shows this calculation for the 12°F profile point.

The ventilation air dampers are closed during the unoccupied period. This means the building will not be pressurized to cut down on infiltration. An average infiltration of 50% of maximum infiltration rate was assumed, to reflect the lower wind velocity at an average condition.

Building electrical input to corridor lights, hot water pump, and fan-coil unit motors is a credit which would reduce boiler power input on a "1 KW for 1 KW" basis. Power input calculated from the block load is, therefore, the total building power input, not the boiler input.

Results of the 52°F profile point calculation and the completed unoccupied period heating profile are shown in **figure SP-37.**

Solar heat gain to the building during the day on weekends and holidays will bring a credit to the heating energy requirement.

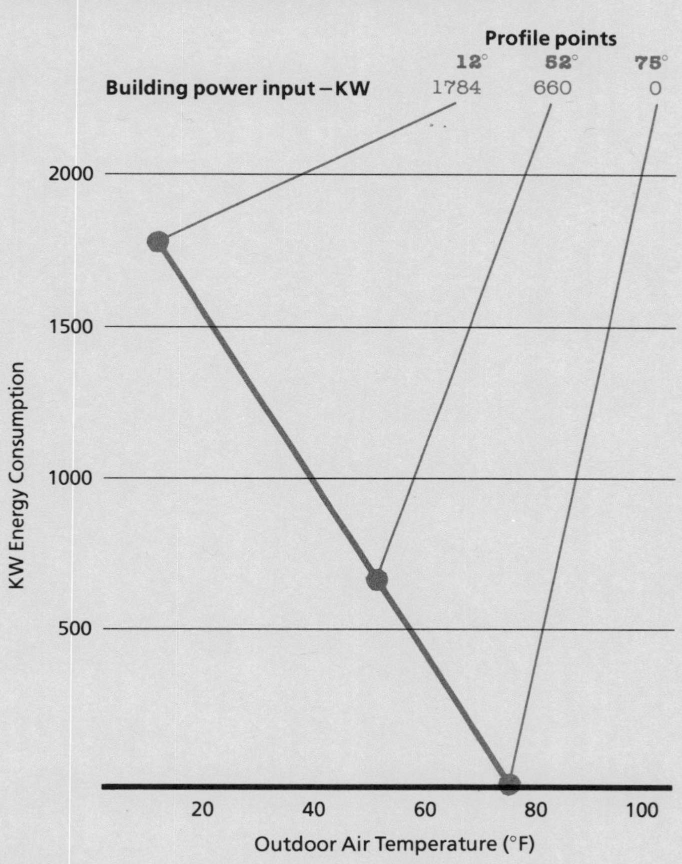

Figure SP-37

Unoccupied Period Heating Profile 12° Outside Temperature

Line	Item	Quantity		Temp. Diff. SHGF		"U" S.C.	Constant	Factors		BTUH
	Losses:									
1	Glass Trans.	65,240 sq. ft.	×	(12°-75°)	×	1.1			=	– 4,521,132
2	Wall Trans.	21,736 sq. ft.	×	(12°-75°)	×	.16			=	– 219,099
3	Pl. Wall Trans.	28,992 sq. ft.	×	(12°-75°)	×	.16			=	– 292,239
4	1st Fl. Trans.	7,764 sq. ft.	×	(12°-75°)	×	.26			=	– 127,174
5	Infiltration	8,435 CFM	×	(12°-75°)	×		1.09		=	– 579,231
6	Machine Rm. Roof	14,628 sq. ft.	×	(12°-52°)	×	.15			=	– 87,768
7	Machine Rm. Walls	8,784 sq. ft.	×	(12°-52°)	×	.75			=	– 263,520
8								**Building Heat Loss**	=	**– 6,090,164**
								÷ 3413 = 1784 KW		

Figure SP-36

This credit is largely independent of outside temperature, so it is easiest to apply after the monthly heating requirements are summarized. How to calculate credit for weekend solar heat gain is shown in **figure SP-38.**

The shade factor used for this credit presumes that building tenants do not open the window blinds.

If the blinds are accurately assumed to be closed when the sun is out, then occupants cannot see out. In many buildings that is what happens. The energy planner should examine how the windows and blinds are used in the sample building to see if this sort of assumption is correct and, if so, what can be done to remedy it.

Step C3 Monthly energy consumption credit

The monthly energy consumption summary consists of separate bin method summaries for each month, for both occupied and unoccupied periods. The summary uses the energy consumption profiles prepared in Steps C1 and C2 and the hours of occurrence for each temperature bin obtained from U.S. Air Force bin weather data.

Figure SP-39 shows the monthly energy consumption summary for January (occupied mode).

Weekend Solar Heat Gain Credit

		Glass area sq. ft.		Day-total SHGF		Correction* Factors		BTU/Day
North solar	=	20,818	×	90	×	.34	=	637,031
East solar	=	13,566	×	421	×	.34	=	1,941,837
South solar	=	17,290	×	1115	×	.34	=	6,554,639
West solar	=	13,566	×	421	×	.34	=	1,941,837
						Solar gain per day	=	11,075,344 BTUH

÷ 3413 = 3245 KWH

÷ 16 Hour operating day = 203 KW

Shade Coeff.		Sunshine Factor		Haze Factor		Dew Pt. Corr.		Sash Corr.		
*.53	×	48%	×	.9	×	1.28	×	1.17	=	0.34

Figure SP-38

Typical Monthly Energy Consumption Summary (Occupied Period)

Month: January **Mode:** Occupied

O.A. °F Temp. Bin	Hours Occurrence	Hourly KW	KWH
97			
92		756	
87		715	
82		675	
77		634	
72	1	593	593
67	8	627	5,016
62	15	661	9,915
57	24	694	16,656
52	33	728	24,024
47	48	820	39,360
42	55	911	50,105
37	56	1,003	56,168
32	47	1,094	51,418
27	27	1,186	32,022
22	17	1,277	21,709
17	7	1,369	9,583
12	5	1,460	7,300
	Monthly energy consumption	=	**323,869 KWH**

Figure SP-39

The unoccupied period summary is similar, except that credit for solar heating of the building on weekend days is deducted from the total. This credit is based on the month's weekend-day hours below 75°F. **Figure SP-40** shows the unoccupied period monthly energy consumption summary for April.

Monthly energy consumption totals for occupied and unoccupied periods are added and shown in **figure SP-41.**

Typical Monthly Energy Consumption Summary (Unoccupied Period)

Month: April **Mode:** Unoccupied

O.A. °F Temp. Bin	Hours Occurrence	Hourly KW	KWH
97			
92			
87			
82			
77			
72	34	86	2,924
67	52	230	11,960
62	74	373	27,602
57	67	517	34,639
52	47	660	31,020
47	42	801	33,642
42	25	941	23,525
37	12	1,082	12,984
32	3	1,222	3,666
27		1,363	
22		1,503	
17		1,644	
12		1,784	

Sub-Total KWH = **181,962**

*Weekend Solar Credit =
126 Hrs. x 203 KW − 25,578

Monthly Energy Consumption = **156,384 KWH**

*Note that this office building uses passive solar heating. The credit can be taken because the temperatures inside can exceed the comfort range: since the building is not occupied, no one will suffer. Also, the cooling system will not try automatically to remove it.

Figure SP-40

Annual Energy Consumption Summary

Month	Bin Method Estimate* Occupied Period	Unoccupied Period	Garage Fans	HVAC Total KWH
January	323,869	384,848	3279	711,996
February	273,179	295,079	2962	571,220
March	283,383	275,115	3279	561,777
April	231,252	156,384	3173	390,809
May	227,642	75,433	3279	306,354
June	220,132	31,464	3173	254,769
July	226,765	16,370	3279	246,414
August	225,945	20,025	3279	249,249
September	218,252	56,300	3173	277,725
October	248,387	147,641	3279	399,307
November	260,483	255,670	3173	519,326
December	292,705	370,742	3279	666,726
Total	**3,032,014**	**2,085,071**	**38,607**	**5,155,672 KWH**

*Includes all HVAC equipment except garage fans

Figure SP-41

Sample Problem

Non-HVAC Energy Calculation

The next move is to make an energy estimate of non-HVAC equipment. The general procedure is the same as before. Parts A (compare equipment capacity with requirement), B (calculate peak energy demand) and C (calculate average energy consumption) are performed for each item. Since the building's non-HVAC equipment does not respond to changes in the weather, the equivalent full-load hours method is accurate enough.

Lighting

Installed lighting capacity can only be compared with actual requirements by a thorough in-field analysis of actual lighting conditions. Every building and, indeed every space in it, has its own requirements as determined by the architecture of the space and the needs of its users.

Light meters can be used to check illumination levels at different locations in the building. These readings are then compared to the recommended task illumination levels. This will tell the energy planner roughly how actual illumination levels compare with requirements. The estimates are rough because meters which measure only footcandles give no indication of light **quality.** This is most important for comfortable vision.

Lighting standards recommended by the Illuminating Engineers Society are always given in terms of equivalent sphere illumination (ESI), a measure of task visibility which also takes into account glare and veiling reflection. Light meters which measure only footcandles do not measure ESI levels. Meters which measure ESI are currently very expensive.

In the end, the simplest and most accurate way to evaluate illumination levels in a building is to interview the users and to note any changes they may have made to improve conditions. Examples include adding their own task lights or shielding their work areas from glare.

The U.S. General Services Administration has published illumination level requirements as part of the Lighting and Thermal Operations Guidelines of the Federal Energy Adminstration.

Another rough approach open to the energy planner is to compare installed lighting wattage per square foot to published standards or to similar buildings **(fig. SP-42).** Note, however, that requirements even for a specific task can vary widely. It takes more capacity to provide light for reading anywhere in a room than it does to provide it in a few specific work areas. Other variations arise from the quality of light, efficiency of the light source and its color characteristics. Reading under one light source may require twice as many watts per square foot as under a light source with better color characteristics.

With this in mind, proceed with the comparison. There is always a good chance that lighting can be sharply reduced yet with an increase in seeing comfort. This opportunity should not be passed up.

The monthly demand and energy consumption estimate for lighting is shown in **figure SP-43.** Lighting

Lighting Capacity vs. Requirements

Area	Tenant Spaces	Parking	Toilets and Stairs	Lobbies** and Corridors	Equipment Room
Installed capacity	4 w/sq. ft.	1 w/sq. ft.	2 w/sq. ft.	2 w/sq. ft.	2 w/sq. ft.
Approximate required capacity*	2 w/sq. ft.	1 w/sq. ft.	1 w/sq. ft.	1 w/sq. ft. corridors only	1 w/sq. ft.

*Less is possible with good lighting design
**Special use areas require individual study

Figure SP-42

for tenant and lobby spaces was combined for use with a common **diversity factor.**

All lighting loads except outside perimeters lighting probably coincide with total building peak demand. Perimeter lighting would not be on at time of peak demand and was therefore tabulated separately.

Elevators

Power input to elevators depends greatly on usage and on equipment characteristics of the installation. It is therefore best either to monitor actual power consumption over a period of time or to bring in the manufacturer of the equipment for a detailed study.

As this is a hypothetical building, assume elevator power input as if monitored over a period of several weeks to evaluate actual building requirements and obtain typical usage figures.

Then compare actual requirements as determined by observation and monitoring of actual use patterns, with installed capacity. This will determine if it is practical to modify the installation and reduce energy consumption.

Peak demand of the 8 elevators was found to be between 8:15 and 8:45 AM; it came to 176 KW. Similar, but lower peaks were recorded between 11:45 AM and 12:45 PM and between 4:45 and 5:15 PM.

Average daily consumption of the 8 elevators was 752 KWH. This was converted to a monthly consumption value by use of a "days usage per month" factor. Monthly usage calculations are shown in **figure SP-44.**

Elevator Power Consumption

Month	*Days Usage per month	KWH Usage per day	KWH per month
January	21.7	752	16,318
February	19.6	752	14,739
March	21.7	752	16,318
April	21.0	752	15,792
May	21.7	752	16,318
June	21.0	752	15,792
July	21.7	752	16,318
August	21.7	752	16,318
September	21.0	752	15,792
October	21.7	752	16,318
November	21.0	752	15,792
December	21.7	752	16,318
			192,135

*Actual days/month × 5/7 days/week × 51/52 weeks/yr.

Figure SP-44

Energy Estimate – Lighting

		Col. (A)	(B)	(A) × (B)	(C)	(D)	(A) × (C)
		Installed	Peak Demand		Average Consumption		
(a) Mon. – Fri.:		KW	Diversity Mult.	KW	Diversity Mult.	Hours Per Day	KWH Per Day
Tenant Spaces		972	.9	875	.8	10	7,774
Interior	501						
Perimeter	429						
Lobbies, Corridor	41.4						
Toilets, Stairs		25.4	1.0	25	1.0	15	381
Parking		75.0	.9	68	.9	15	1,013
Equipment Room		30.0	.9	27	.9	16	432
			Totals	995			9,600
(b) 7 Days Per Week:							
Perimeter		5.1	Not Coincident		1.0	4	20

	Feb 28 Days	Apr, June Sept, Nov 30 Days	Other Months 31 Days
(a) 9600 KWH/day × days/mo × 5/7 days/week × 51/52 weeks/yr.	188,308	201,758	208,484
(b) 20 KWH/day × days/month	560	600	620
Lighting KWH/month	**188,868**	**202,358**	**209,104**

Figure SP-43

Sample Problem

Domestic hot water

The domestic hot water heater capacity is 100 KW, and it is provided with a 1,000 gallon storage tank. At present, hot water supply temperature is controlled at 140°F.

Actual requirements for domestic hot water in an office building are minimal.

For example, assuming that local high temperature requirements (such as for a snack bar dishwasher water) are met by a local booster heater, the restroom lavatory requirements can be satisfied with much lower temperature. Assume 80°F as a minimum requirement for our estimate of required capacity.

The amount of water used for washing hands does not vary appreciably with supply temperature. Average daily usage based on an assumed 1.5 gallons per person is:

1,760 persons @ 1.5 gal.	2,640
Allowance for janitor closets and leakage	360
Total estimated use	3,000 gal./day

Power input required to raise water temperature from 45°F annual minimum temperature to 80°F is:

$$3{,}000\,\frac{\text{GAL}}{\text{DAY}} \times 8.33\,\frac{\text{LBS}}{\text{GAL}} \times (80°\text{F}-45°\text{F}) \times \frac{1\ \text{BTU}}{°\text{F-LB}}$$

$$\div\ 3{,}413\,\frac{\text{BTU}}{\text{KWH}} = 256\ \text{KWH}$$

Domestic Hot Water Heating

Note	(1)		(2)	(3)	(4)	(5)	(6)
Month	**Entering Water Temp**	**Temp. Diff.**	**KWH per day**	**Days Use per month**	**KWH per month**	**KWH Insul. Loss per month**	**Total KWH per month**
January	45	140-45 = 95°	696	21.7	15,103	4309	19,412
February	47	140-47 = 93°	681	19.6	13,343	3892	17,235
March	54	140-54 = 86°	630	21.7	13,661	4309	17,970
April	61	140-61 = 79°	578	21.0	12,144	4170	16,314
May	64	140-64 = 76°	556	21.7	12,072	4309	16,381
June	70	140-70 = 70°	512	21.0	10,760	4170	14,930
July	74	140-74 = 66°	483	21.7	10,484	4309	14,793
August	70	140-70 = 70°	512	21.7	11,119	4309	15,428
September	70	140-70 = 70°	512	21.0	10,760	4170	14,930
October	62	140-62 = 78°	571	21.7	12,390	4309	16,699
November	60	140-60 = 80°	586	21.0	12,298	4170	16,468
December	53	140-53 = 87°	637	21.7	13,819	4309	18,128
							198,688

Notes:
(1) from local water department
(2) 3000 gal/day × 8.33#/gal × temp diff ÷ 3413 BTU/KWH
(3) Actual days × 5/7 days/week × 51/52 weeks/year
(4) Columns (2) × (3)
(5) Tank Area = 4′ Dia. × 10′ Long × TT = 126 sq. ft. × .35 "U" × (140-75°) = 2,867 BTUH
Piping loss = 1000 ft. × .26 BTUH/foot × (140-75) = 16,900 BTUH
System Heat Loss = 19,767 BTUH
÷ 3413 BTU/KWH × 24 hrs/day = 139 KWH/day
× Actual Number days/month = (5)
(6) Sum of (4) and (5)

Figure SP-45

Assuming a 10 hour period for heater operation, a 25 KW heater would do the job. An even smaller heater would do if tank temperature were maintained at 180°F, and a mixing valve were provided at the tank outlet to mix tank water and cold water to obtain the required 80°F supply water.

The 100 KW heater can operate to coincide with peak HVAC and elevator use. Peak demand must, therefore, be considered as 100 KW.

Average energy consumption was estimated based on maintaining the present 140°F hot water supply temperature. Water usage is the same 3,000 gallons per day estimated before. **Figure SP-45** shows the completed monthly domestic hot water heating estimate.

House pumps

The plumbing system is served from a house tank located on the 20th floor. All domestic water for toilets and lavatories must be pumped to this house tank, where it flows by gravity to the point of use.

Power input to the house pumps was obtained by measuring power input, and by monitoring the house pump operation with a timer to obtain actual hours of operation for a typical day. One pump operated 6.8 hours on an average day.

The house tank is large enough to allow the pump to operate continuously during normal occupancy hours, so the pump has considerable excess capacity. Considering the small size of the pump, however, it is not practical to consider changing pump capacity to more closely match the actual load.

A second pump was arranged for stand-by operation through a transfer switch, so both pumps would never operate simultaneously. Peak demand therefore consists of the kilowatts of one pump in operation.

Peak demand and average energy consumption estimates are shown in **figure SP-46.**

House Pumps

Each pump: Peak demand (only one pump operates)

$$\text{KW Input} = \frac{7.1 \text{ Amps} \times 467 \text{ Volts} \times 1.73 \times .8 \text{ PF}}{1000} = 4.6 \text{ KW}$$

× 6.8 Avg. hours operation = **31.2 KWH per occupied day**

Monthly Consumption:

February : 31.2 KWH × 5/7 days/wk. × 51/52 weeks/yr. × 28 days = **612 KWH**
30 Day Months : 31.2 KWH × 5/7 days/wk. × 51/52 weeks/yr. × 30 days = **656 KWH**
31 Day Months : 31.2 KWH × 5/7 days/wk. × 51/52 weeks/yr. × 31 days = **678 KWH**

Figure SP-46

Office equipment

Assume that miscellaneous installed office equipment totals 0.5 watts per square foot of office area. This is standard for buildings where such equipment accounts for a small portion of total energy use. Even so, comparing installed capacities to requirements for a building of this type is best done on site.

The owner and building users can be questioned as to the amount and size of equipment needed and daily periods when actually used. A building will often contain several coffee pots or copy machines which are left "on" all day even though used only for one or two hours. Identifying such information is a good example of a task that can be done equally well by the owner and his task force or by the energy planner. More important are computers and other items of large energy-consuming business equipment. These call for a careful inventory.

The monthly demand and energy consumption estimate for office equipment is shown in **figure SP-47.** It was assumed that during peak periods, only 50% of all equipment would be operating; during average periods, only 25%.

Energy Estimate – Office Equipment

Peak Demand:

116 KW Installed Capacity × .5 diversity = 58 KW

Average Energy Consumption:

116 KW × .25 Avg. diversity × 8 Hours/Day = **232 KWH/day**

Monthly Consumption:

February	: 232 KWH × 5/7 days/wk. × 51/52 weeks/yr. × 28 days	=	**4551 KWH**
30 Day Months	: 232 KWH × 5/7 days/wk. × 51/52 weeks/yr. × 30 days	=	**4876 KWH**
31 Day Months	: 232 KWH × 5/7 days/wk. × 51/52 weeks/yr. × 31 days	=	**5038 KWH**

Figure SP-47

Annual energy summary

The next task is to total monthly peak energy demand and average energy consumption figures for HVAC and Non-HVAC equipment.

Annual Total Energy Summary

Month	HVAC	Lighting	Elevators	Domestic H.W.	House Pumps	Office Equip.	Total
			Peak Demand (KW)				
January	2,233	995	176	100	4.6	58	**3566.6**
February	2,233	995	176	100	4.6	58	**3566.6**
March	2,233	995	176	100	4.6	58	**3566.6**
April	1,793	995	176	100	4.6	58	**3126.6**
May	1,387	995	176	100	4.6	58	**2720.6**
June	1,103	995	176	100	4.6	58	**2436.6**
July	1,103	995	176	100	4.6	58	**2436.6**
August	1,103	995	176	100	4.6	58	**2436.6**
September	1,184	995	176	100	4.6	58	**2517.6**
October	1,996	995	176	100	4.6	58	**3329.6**
November	2,233	995	176	100	4.6	58	**3566.6**
December	2,233	995	176	100	4.6	58	**3566.6**
			Consumption (KWH)				
January	711,996	209,104	16,318	19,412	678	5,038	**962,546**
February	571,220	188,868	14,739	17,235	612	4,551	**797,225**
March	561,777	209,104	16,318	17,970	678	5,038	**810,885**
April	390,809	202,358	15,792	16,314	656	4,876	**630,805**
May	306,354	209,104	16,318	16,381	678	5,038	**553,873**
June	254,769	202,358	15,792	14,930	656	4,876	**493,381**
July	246,414	209,104	16,318	14,793	678	5,038	**492,345**
August	249,249	209,104	16,318	15,428	678	5,038	**495,815**
September	277,725	202,358	15,792	14,930	656	4,876	**516,337**
October	399,307	209,104	16,318	16,699	678	5,038	**647,144**
November	519,326	202,358	15,792	16,468	656	4,876	**759,476**
December	666,726	209,104	16,318	18,128	678	5,038	**915,992**
Total	**5,155,672**	**2,462,028**	**192,133**	**198,688**	**7,982**	**59,321**	**8,075,824**

Annual consumption/S.F. = 26 KWH/S.F. or 89 MBTU/S.F.

Figure SP-48

Energy cost calculations

Finally, total monthly electric cost is estimated by applying the previous peak demand and average consumption totals to the Georgia Power Company's PL-1 rate schedule. **Figure SP-49** shows the electric cost calculations for January using the actual rate schedule.

Figure SP-50 shows the January calculations using Georgia Power Company's short-cut formula. This formula is far easier to use and provides a fair estimate of the monthly demand charge, even though demand and consumption charges are lumped together in the actual rate.

Figure SP-51 shows the monthly electric costs and annual total, calculated using the short-cut formula. Note that the results of the two calculations match.

Electricity Cost Calculations (using schedule "PL-1" short cut formula)

Month: January

2,392 KW-Billing Demand 962,546 KWH

962,546 KWH / 2,392 KW = 402 Hours use of demand

	From Proper table (1), (2), or (3)		
Correction Factor		$	1,317.25
Demand Charge (if any)	(2,392 KW × 5.42 $/KW)	$	12,964.64
Energy Charge	(962,546 KWH × .013 $/KWH)	$	12,513.10
	(Base Bill) Subtotal	$	26,794.99
+ Fuel Adjustment	(962,546 KWH × .00227 $/KWH)	$	2,184.98
	Subtotal	$	28,979.97
+ Sales Tax (4%)		$	1,159.20
	Billing Total	**$**	**30,139.17**

Figure SP-50

Electricity Cost Calculations (using rate schedule "PL-1")

Month January **Consumption** 962.546 KWH **Demand** 3566.6 KWH

(1) Calculate billing demand

actual demand	3566.6
.95 × high summer demand	2392
.60 × high winter demand	2140
(Include current month in winter)	therefore use 2392 KW

(2) Calculate minimum bill

$5.00 + $3.50 [(1) − 5KW = 2387] = $ 8,359.50 plus fuel adjustment and sales tax

(3) Calculate energy charge including demand charge:

z = billing demand 2392 × 200 hrs = 478,400

y = billing demand 2392 × 400 hrs = 956,800

(z > consumption > y)	478,400 × $.015		= $ 7,176.00
(consumption > z)	5,746 × $.013		= $ 74.70
(1-50 KWH)	amount 50 × $.100	= 5.00	
(51-1500 KWH)	amount 1,450 × $.065	= 94.25	
(1501-3000 KWH)	amount 1,500 × $.062	= 93.00	
(3001-10,000 KWH)	amount 7,000 × $.055	= 385.00	
(10,001-200,000 KWH)	amount 190,000 × $.044	= 8,360.00	
(200,001-500,000 KWH)	amount 287,340 × $.0381	= 10,947.65	
(500,001 KWH +)	amount ______ × $.035	= ______	
		Total	= **$ 27,134.25**

(4) Calculate Fuel Adjustment:

Consumption 962,546 × $.00227/KWH = **$ 2,184.98**

(5) Calculate total utility charge: (3) + (4) = **$ 29,319.23**

(6) Calculate state sales tax: (5) × .04 = **$ 1,172.77**

(7) Calculate total bill: (6) + (5) = **$ 30,492.00**

Figure SP-49

Average Annual Electric Costs (using Schedule "PL-1" short cut formula)

Total Annual Costs = $300,577.64 = $.94/gross sq. ft.

Month	(1) Consumption (KWH)	(2) Peak Demand (KW)	(3) Billing Demand (KW)	(4) Hrs Use	(5) Correction Factor	(6) Demand* Charge	(7) Consumption* Charge	(8) Fuel Adj.	(9) Tax	(10) Total
January	962,546	3567	2392	402	$1,317.25	$12,964.64	$12,513.10	$2,184.98	$1,159.20	$30,139.17
February	797,225	3567	2392	333	1,317.25	11,051.04	11,958.38	1,809.70	1,045.46	27,181.83
March	810,885	3567	2392	339	1,317.25	11,051.04	12,163.28	1,804.71	1,053.45	27,389.73
April	630,805	3127	2392	264	1,317.25	11,051.04	9,462.08	1,431.93	930.49	24,192.79
May	553,873	2721	2392	232	1,317.25	11,051.04	8,308.10	1,257.29	877.35	22,811.03
June	493,381	2437	2437	202	1,317.25	11,258.94	7,400.72	1,119.97	843.88	21,920.76
July	492,345	2437	2437	202	1,317.25	11,258.94	7,385.18	1,117.62	843.16	21,922.15
August	495,815	2437	2437	203	1,317.25	11,258.94	7,437.23	1,125.50	845.57	21,984.49
September	516,337	2518	2518	205	2,867.25	10,072.00	7,745.06	1,172.08	874.26	22,730.65
October	647,144	3330	2392	271	1,317.25	11,051.04	9,707.16	1,469.02	941.78	24,486.25
November	759,476	3567	2392	318	1,317.25	11,051.04	11,392.14	1,724.01	1,019.38	26,503.82
December	915,992	3567	2392	383	1,317.25	11,051.04	13,739.88	2,079.30	1,127.50	29,314.97
Total	**8,075,824**					**$134,170.74**	**$136,549.31**	**$18,296.11**	**$11,561.48**	**$300,577.64**

4: (1) ÷ (3)
5: From short cut formula
6: (3) × $/KW from short cut formula ($4.62/KW except September & January) Tables 2 & 3
7: (1) × $.015 from short cut formula Table 2 (January = 0) × $.013 – Table 3
8: (1) × $.00227 (estimated fuel adjustment from electric co.)
9: [(5) + (6) + (7) + (8)] × .04
10: (5) + (6) + (7) + (8) + (9)

$$\frac{\text{CONSUMPTION + FUEL ADJ.}}{\text{DEMAND + CONSUMPTION + FUEL ADJ.}} = 54\%$$

*Note that the "demand charge" in many instances exceeds the "consumption charge."

Figure SP-51

Graphic analysis

Calculating average monthly energy costs completes the sample problem. The following charts and graphs have been generated from sample problem figures. They are an example of the kind of graphic analysis which serves to identify and evaluate opportunities for improved energy performance.

A thorough evaluation would involve changing input data to reflect the impact of any modifications being considered, then reworking the energy estimates to produce new peak demand, average consumption, and cost figures.

Changes in required capacity should also be studied, since any modification which, say, reduces HVAC loads, will also reduce HVAC capacity requirements. The new figures can then be compared to the original results to produce an accurate estimate of the change to expect in the building's energy performance.

Figure SP-52 shows a breakdown of September and January peak electric demand.

Figure SP-53 shows a plot of monthly actual and billing peak electric demand figures.

Figure SP-54 shows a breakdown of the average annual electric consumption. HVAC and lighting consumption have been individually broken down to show the contribution of each.

Figure SP-55 shows a plot of the monthly electric consumption.

The following figures show the information needed to compile the breakdown of HVAC and lighting consumption components as shown in **figure SP-54.** This information was based entirely on figures SP-1 through SP-51 so the reader can more easily reconstruct it.

Not all the calculations needed to complete the sample problem were reproduced. As a result, some approximations were used to prepare the following figures. This has led to slightly different totals than those in the sample problem (figs. SP-1 through SP-51). The results are, however, accurate enough for a quick evaluation of modifications, and that is what they are intended to illustrate.

To prepare the breakdown of HVAC and lighting consumption shown in **figure SP-54,** each had to be broken down into its various components. For HVAC consumption, a breakdown of both occupied and unoccupied consumption was required. **Figure SP-56** shows the lighting consumption breakdown.

Peak Electric Demand Breakdown from figure SP-48

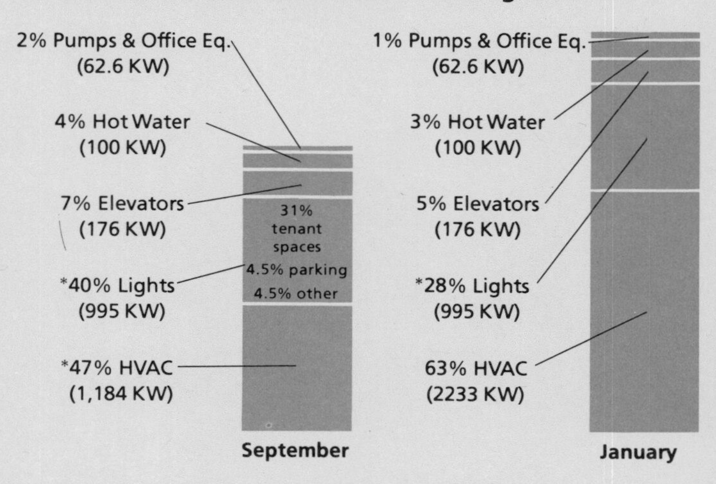

*Consider the significance of lights also as a cause of increased cooling and a contributor to heat.

Figure SP-52

Monthly Peak Electric Demand from figure SP-48

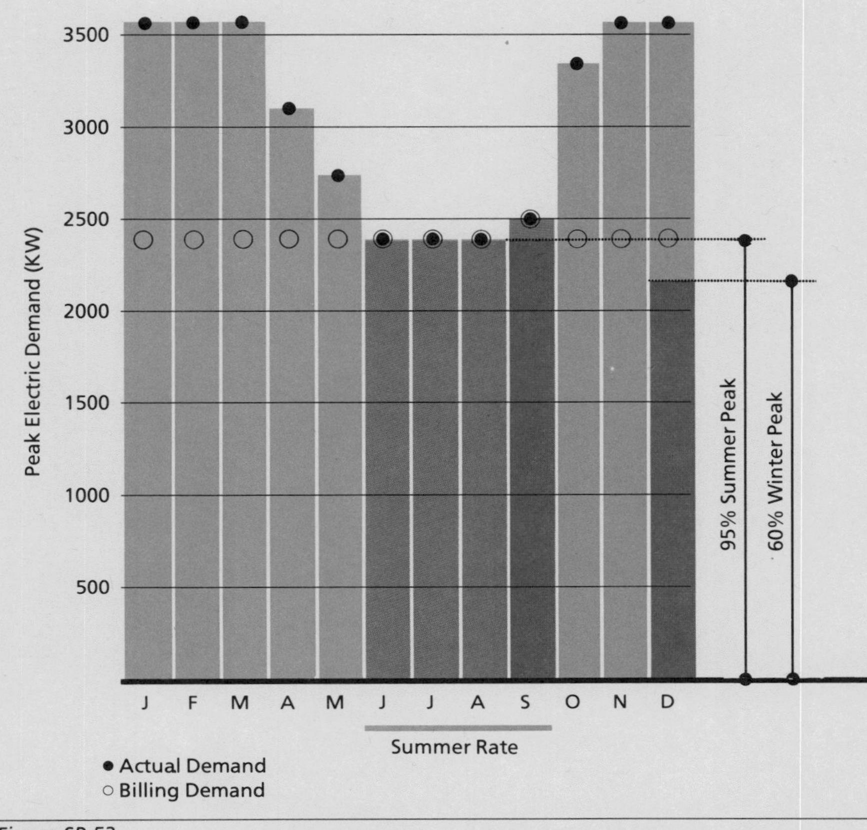

Figure SP-53

Average Annual Electric Consumption Breakdown from figure SP-48

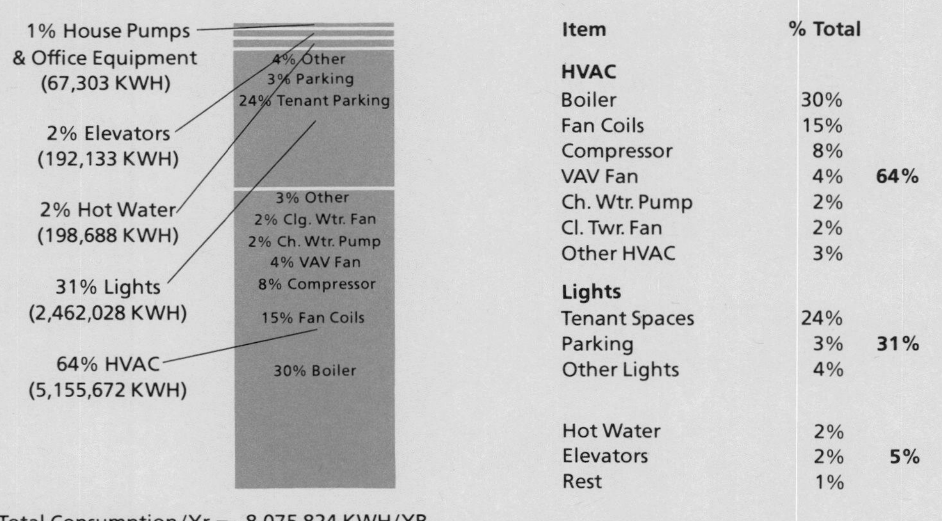

Item	% Total	
HVAC		
Boiler	30%	
Fan Coils	15%	
Compressor	8%	
VAV Fan	4%	**64%**
Ch. Wtr. Pump	2%	
Cl. Twr. Fan	2%	
Other HVAC	3%	
Lights		
Tenant Spaces	24%	
Parking	3%	**31%**
Other Lights	4%	
Hot Water	2%	
Elevators	2%	**5%**
Rest	1%	

Total Consumption/Yr = 8,075,824 KWH/YR
= 27,562,787 MBTU/Yr = 86 MBTU/gross sq. ft./yr.

Figure SP-54

Figure SP-58 shows the graphic breakdown of HVAC consumption – occupied, unoccupied and total.

To generate the HVAC consumption breakdown, consumption of the HVAC components at each temperature bin was determined and multiplied by the total number of hours that temperatures in each bin occurred. This called for drawing a separate profile for each of the HVAC components.

Figures SP-59 through SP-63 show the HVAC consumption summary by bin for occupied hours, and the individual profiles required to generate it. **Figures SP-64 and SP-65** show the same information for the unoccupied hours.

Figure SP-66 shows the annual summary of HVAC consumption by bin. **Figures SP-67, 68 and 69** are plots of this information. These plots can be used for identifying the outside temperatures at which most energy is consumed and where, consequently, most of the opportunities lie.

Monthly Electric Consumption from figure SP-48

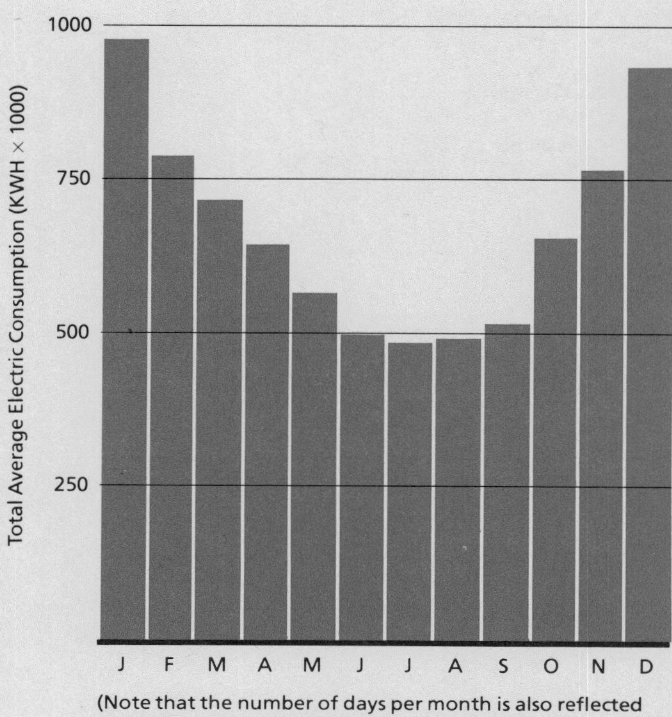

(Note that the number of days per month is also reflected in this chart; January had 31 days, February 28).

Figure SP-55

Average Annual Lighting Energy Consumption Breakdown

Item	% Total
Tenant Spaces	78%
Parking	11%
Toilets & Stairs	4%
Equipment Room	4%
Lobbies & Corridors	3%
Perimeter	–
total Av. KWH/Yr.	2,462,028

Figure SP-56

Average Annual HVAC Energy Consumption Breakdown

Item	% Occupied	% Unoccupied	% Total
Boiler	28%	70%	46%
Fan Coils	21%	28%	24%
Compressor	22%		13%
VAV Fan	11%		6%
Ch. Wtr. Pump	6%		4%
Cl. Twr. Fan	5%		3%
Chd. Wtr. Pump	4%		2%
Ht. Wtr. Pump	1%	2%	1%
Exhaust Fans	2%		1%
Total Av. KWH/Yr.	3,046,546	2,074,336	5,120,882

Figure SP-57

Average Annual HVAC Energy Consumption Breakdown

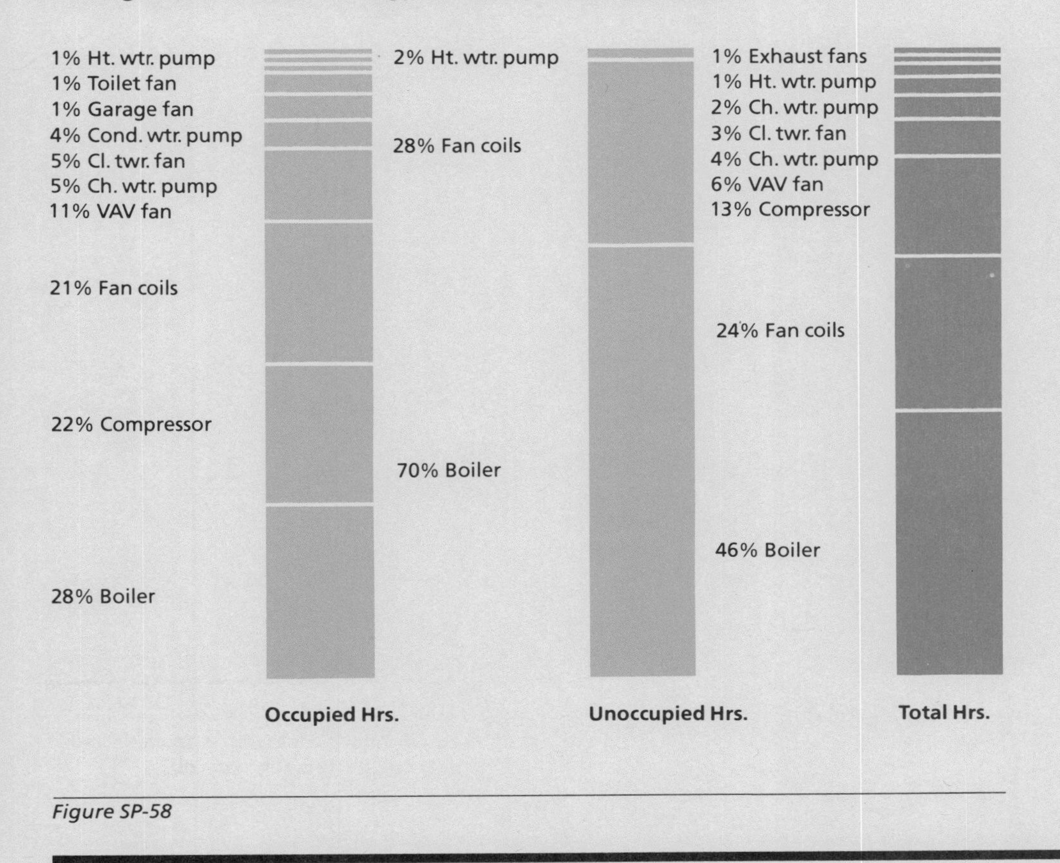

Figure SP-58

Average Annual HVAC Energy Consumption Breakdown – Occupied Hours
KW from individual profiles taken from Figure SP-35

		Fan Coils		VAV Fan		Ch. Wtr. Pump		Compressor		Cnd. Wtr. Pump		Cl. Tw. Fan		Hot Wtr. Pump		Boiler		Toilet Ex. Fan		Eq. Rm. Fan		Garage Fan	
Bin	Hrs	KW	KWH	KW	KWH	KW	KWH	KW	KWH	KW	KWH	KW	KWH	KW	KWH	KW	KWH	KW	KWH	KW	KWH	KW	KWH
92°	77	153	11781	78	6006	48	3696	353	27181	36	2772	75	5775					6	462	7	539	10	770
87°	244	153	37332	78	19032	48	11712	320	78080	36	8784	70	17080					6	1464	7	1708	10	2440
82°	345	153	52785	78	26910	48	16560	285	98325	36	12420	64	22080					6	2070	7	2415	10	3450
77°	382	153	58446	78	29796	48	18336	250	95500	36	13752	59	22538					6	2292	7	2674	10	3820
*72°	497	153	76041	78	38766	48	23856	218	108346	36	17892	54	26838					6	2982			10	4970
67°	434	153	66402	78	33852	48	20832	180	78120	33	14322	49	21266	10	4340			6	2604			10	4340
62°	368	153	56304	78	28704	48	17664	148	54464	30	11040	44	16192	10	3680			6	2208			10	3680
57°	338	153	50949	78	25974	48	15984	120	39960	28	9324	38	12654	10	3330	220	73260	6	1998			10	3330
52°	315	153	48195	78	24570	48	15120	88	27720	25	7875			10	3150	320	100800	6	1890			10	3150
47°	306	153	46818	78	23868	48	14688	75	22950	21	6426			10	3060	420	128520	6	1836			10	3060
42°	271	153	41463	78	21138	48	13008	63	17073	18	4878			10	2710	530	143630	6	1626			10	2710
37°	222	153	33966	78	17316	48	10656	52	11544	14	3108			10	2220	650	144300	6	1332			10	2220
32°	157	153	24021	78	12246	48	7536	40	6280	11	1727			10	1570		117436	6	942			10	1570
27°	78	153	11934	78	6084	48	3744	30	2340	7	546			10	780	850	66300	6	468			10	780
22°	39	153	5967	78	3042	48	1872	20	780	4	156			10	390	960	37440	6	234			10	390
17°	14	153	2142	78	1092	48	672	10	140					10	140	1060	14840	6	84			10	140
12°	10	153	1530	78	780	48	480							10	100	1160	11600	6	60			10	100
	4092	626,076		319,176		196,416		668,803		115,022		144,423		25,470		838,126		24,552		7,336		40,920	
% Total		21%		11%		6%		22%		4%		5%		1%		28%		1%		—		1%	

*Note that most hours occur at the 72° temperature bin as does the maximum energy consumption while this temperature is ideally within the human comfort range. Also, of the energy used at this bin, only the compressor is responding to the weather. The reader can draw his own conclusions.

Figure SP-59

Sample Problem

Compressor: Average Energy Profile – Occupied Hours, from figure SP-35

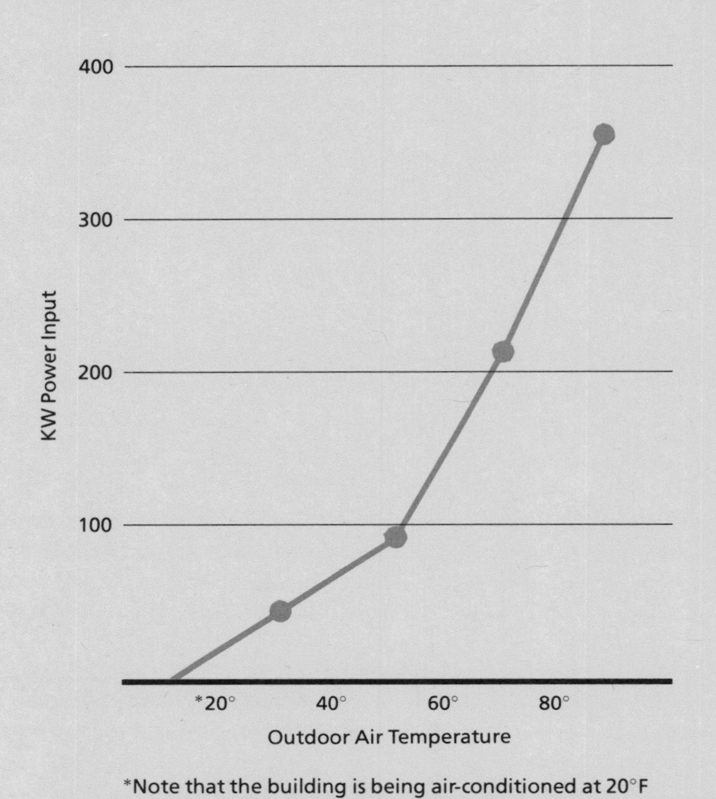

*Note that the building is being air-conditioned at 20°F to melt the cooling load caused by solar heat gain when the solar load exceeds the heat loss at the perimeter. One might consider circulating chilled water to the fan coils through an outside air stream to create a "water side economizer" cycle.

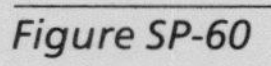

Figure SP-60

Condenser Water Pump: Average Energy Profile – Occupied Hours from figure SP-35

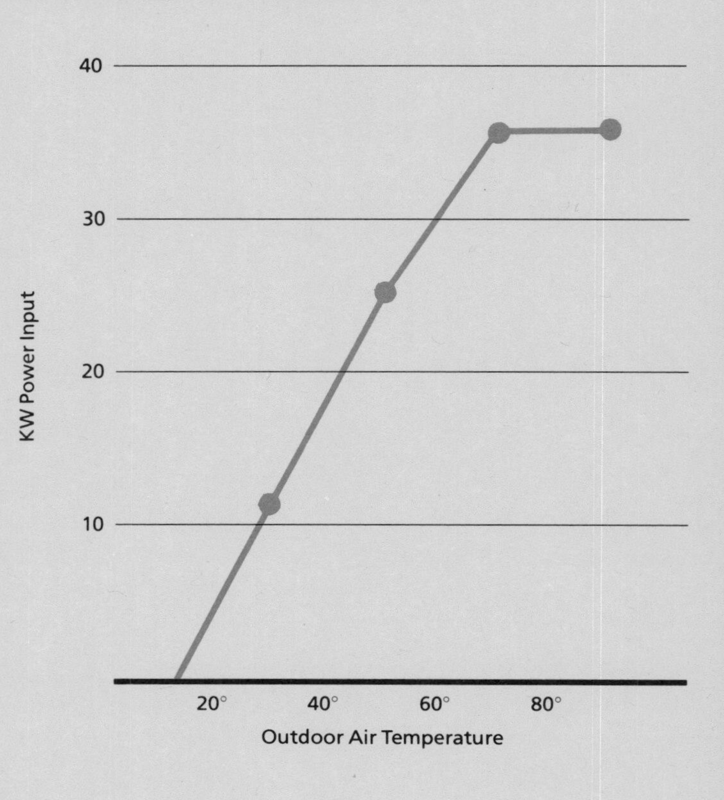

Figure SP-61

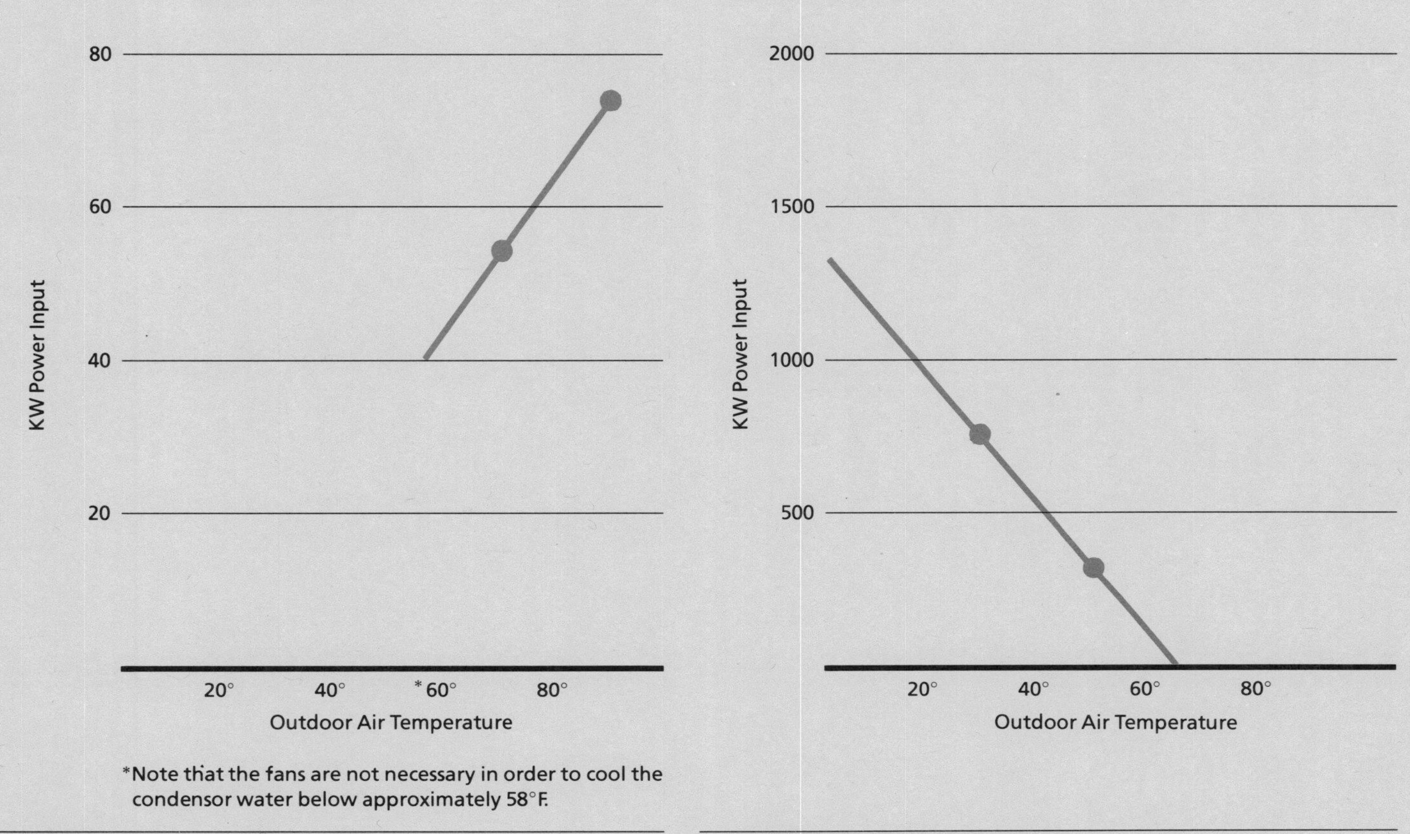

Figure SP-62

Figure SP-63

Average Annual HVAC Energy Consumption Breakdown Unoccupied Hours

		Fan Coils		Hot Wtr. Pump		Boiler			Avg. KWH
Bin	Hrs	KW	KWH	KW	KWH	KW	Avg. KWH*	Solar −Credit	Solar Credit
92°									
87°									
82°									
77°									
72°									
67°	553	153		10		67	37051	− 37758	—
62°	444	153		10		210	93240	− 32074	61166
57°	397	153		10		354	140538	− 29029	111509
52°	369	153		10		497	183393	− 27405	155988
47°	375	153		10		638	239250	− 26593	212657
42°	344	153		10		778	267632	− 23548	244084
37°	303	153		10		919	278457	− 19285	259172
32°	229	153		10		1059	242511	− 13601	228910
27°	118	153		10		1200	141600	− 6902	134698
22°	60	153		10		1340	80400	− 3248	77152
17°	20	153		10		1481	29620	− 1218	28402
12°	15	153		10		1621	24315	− 812	23503
T	**4028**	**153**	**616284**	**10**	**40280**		**1,758,007**		**1,537,241**
% Total			28%		2%				70%

A=1,273,486

A+B = 1,981,432 (B FROM SP-59)

*(Boiler KW from Figure SP-40 less fan coil KW and hot water pump KW)

Figure SP-64

Boiler: Average Energy Profile – Unoccupied Hours from figure SP-40 minus fan coil KW & hot water pump KW

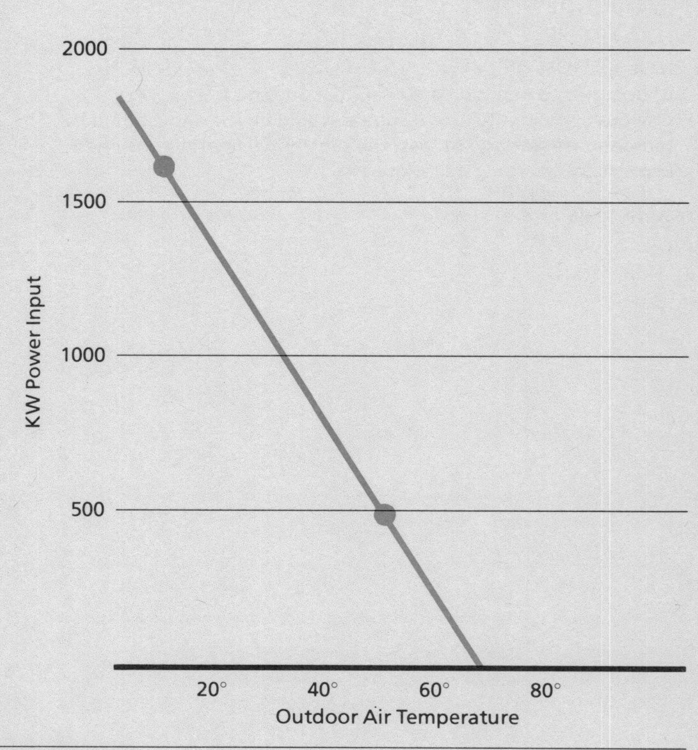

Figure SP-65

Average Annual HVAC Electric Consumption

	(1)	(2)	(3)	(4)	(5)	(6)	(7)	(8)	(9)
	Occupied Hours					**Unoccupied Hours**			
O.A. °F Temp. Bin	**Hours**	**Av. KW**	**Av. KWH**	**Hours**	**Av. KW**	**Av. KWH**	**Weekend Solar Credit**	**Total Av. KWH**	**Total Av. KWH**
92°	77	756	58,212	41					58,212
87°	244	715	174,460	140					174,460
82°	345	675	232,875	241					232,875
77°	382	634	242,188	379					242,188
72°	497	593	294,721	640	86	55,040	43,239	11,801	306,522
67°	434	627	272,118	553	230	127,190	37,758	89,432	361,550
62°	368	661	243,248	444	373	165,612	32,074	133,538	376,786
57°	333	694	231,102	397	517	205,249	29,029	176,220	407,322
52°	315	728	229,320	369	660	243,540	27,405	216,135	445,455
47°	306	820	250,920	375	801	300,375	26,593	273,782	524,702
42°	271	911	246,881	344	941	323,704	23,548	300,156	547,037
37°	222	1,003	222,666	303	1,082	327,846	19,285	308,561	531,227
32°	157	1,094	171,758	229	1,222	279,838	13,601	266,237	437,995
27°	78	1,186	92,508	118	1,363	160,834	6,902	153,932	246,440
22°	39	1,277	49,803	60	1,503	90,180	3,248	86,932	136,735
17°	14	1,369	19,166	20	1,644	32,880	1,218	31,662	50,828
12°	10	1,4600	14,600	15	1,784	26,760	812	25,948	40,548
Total	**4,092**		**3,046,546**	**4,668**		**2,339,048**	**264,712**	**2,074,336**	**5,120,882**

1 & 4: from Temperature Bin Breakdown as in Figure SP-8
2: From Figure SP-39
3: (1) × (2)
5: From Figure SP-40
6: (4) × (5)
7: (Weekend Hours from Temperature Bin Breakdown) × (203 KW/Hr. Solar Credit – Figure SP-40)
8: (6) – (7)
9: (3) + (8)

*Note that 40% of the energy is consumed when the building is unoccupied to maintain a 70°F indoor air temperature. An obvious opportunity to investigate would be the impact of night temperature setback. At least the substantial energy consumed in distribution would be reduced.

Figure SP-66

HVAC Annual Consumption – Occupied Hours

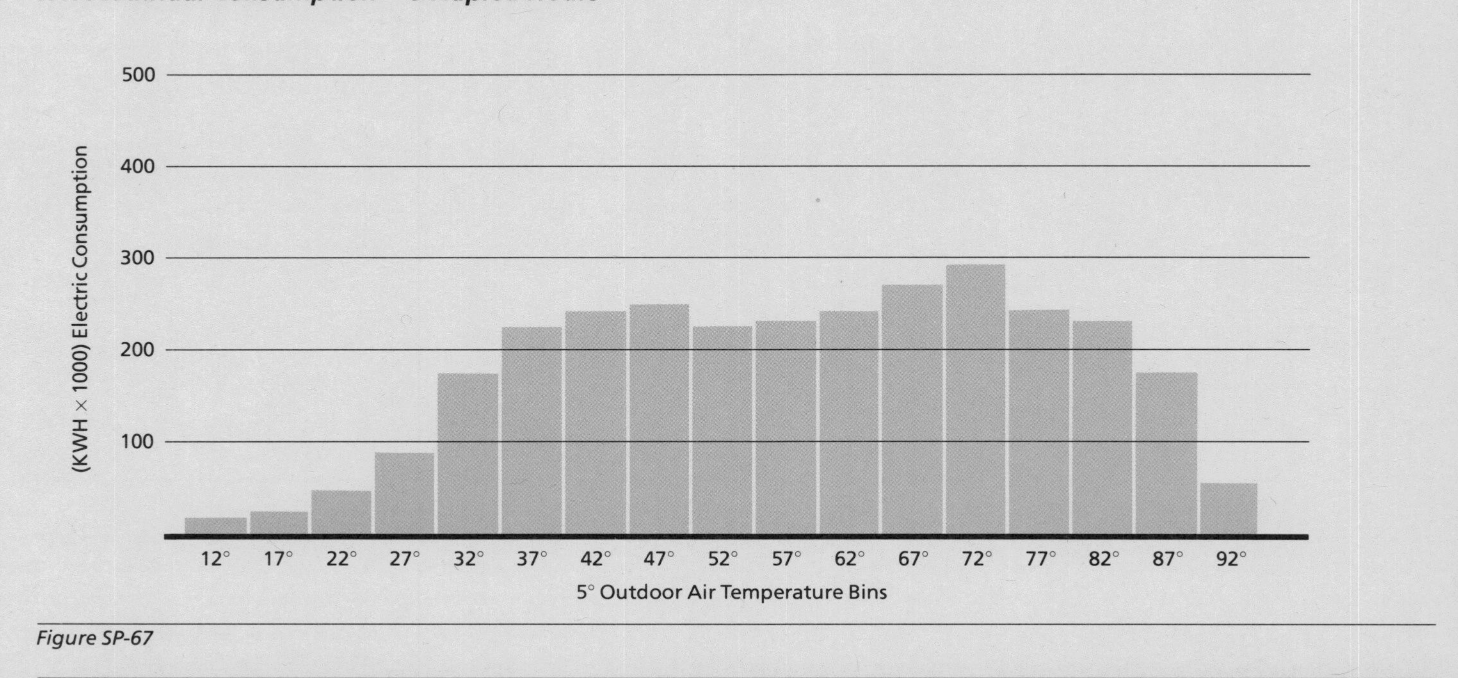

Figure SP-67

HVAC Annual Consumption – Unoccupied Hours

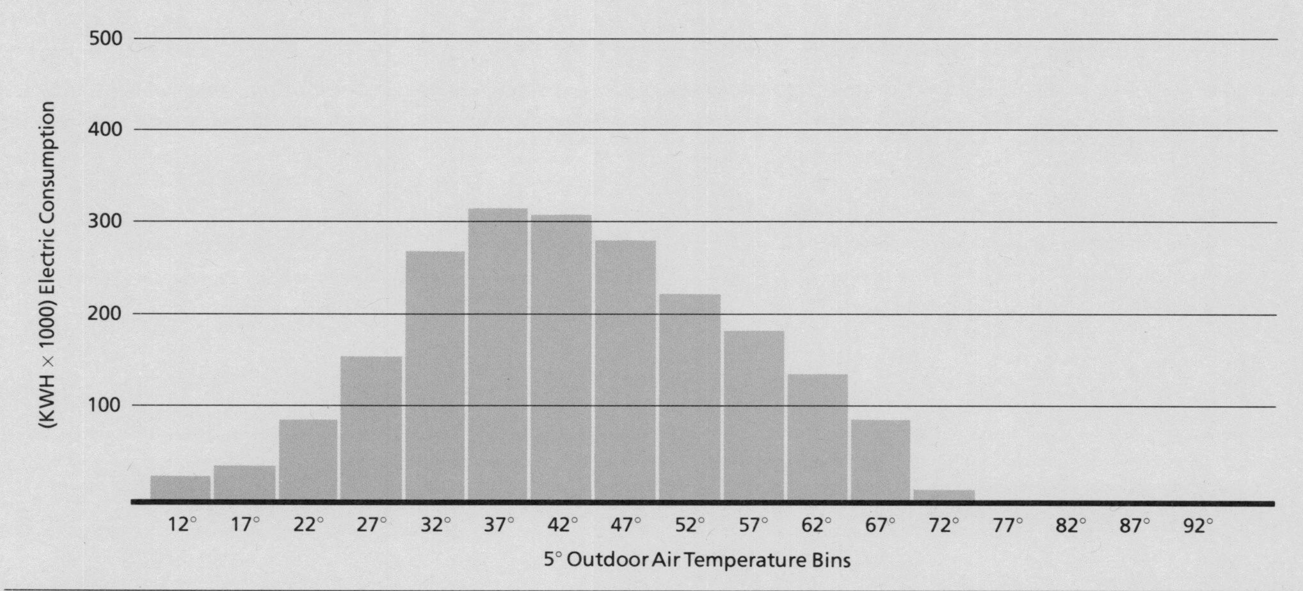

Figure SP-68

HVAC Annual Consumption – Total Hours

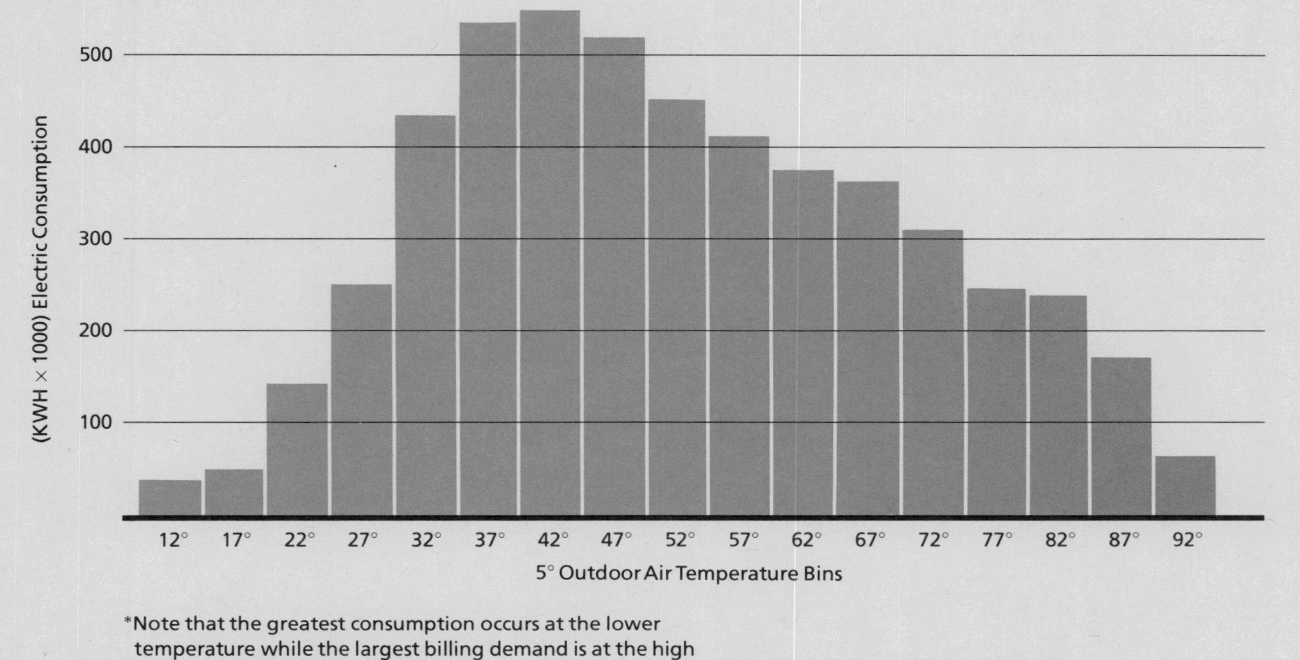

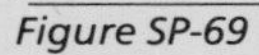

Figure SP-69

Appendix

Appendix A: System Response

Discussion of the bin method pointed out that the system response step is often performed wrongly or overlooked entirely, with major errors in energy calculation. System response is the most complicated part of HVAC energy analysis because there are so many (20 or more) types of HVAC systems and each usually may be arranged or controlled in many different ways. This section provides greater detail on this highly technical subject than would have been reasonable when describing the overall energy estimating method.

The system response step is aimed at relating building thermal load to loads imposed on the purchased energy consuming equipment by a specific type of HVAC system arranged and controlled in a specific manner. Do not, therefore, try to calculate system response without first identifying the system type, arrangement and control.

System response applies mainly to central-type HVAC systems. With a unitary system, the load on zone equipment is usually the zone cooling or heating load.

System response may be broken down into two parts: zone system response and central system response.

Zone system response includes these general items:

- Adding fan motor heat from local zone equipment if not originally included in the zone thermal load estimate. If zone fan motors operate continuously, they provide a constant zone cooling load.
- Determining air and water temperatures supplied to zone equipment at this profile point operating condition, and adjusting available zone equipment cooling or heating capacity accordingly.
- Simulating the action of zone temperature controls to balance adjusted zone thermal load to adjusted zone equipment capacity. This point determines the degree of reheat or mixing of cold and warm air.
- Determining the air and water flows required from fans and pumps, as well as cooling and heating energy required from cooling and heating equipment. These equipment loads are posted to either the zone or central equipment, as determined by the system arrangement.

The action of zone temperature controls of any type of system is defined by examining how the system is arranged and how the zone thermostat responds to a change in load. Zone control is either "sequenced," where cooling capacity and heating capacity are sequenced so both cannot operate at the same time; or "reheat," where cooling and heating energy are used at the same time. Reheat is obviously wasteful, but it is often the most practical way to provide required zone control over temperature and humidity. Reheat is used in many different ways, and not all are equally wasteful. If the system uses reheat for control, one modification objective is to reduce or eliminate this reheat.

Examples of sequenced systems are those using individual self-contained units or chilled water type fan-coil units. The purest example of reheat is the constant volume terminal reheat system.

Many systems use a combination of control methods. Constant volume dual-duct and conventional multi-zone systems are usually sequenced in the cooling season and of the reheat type in the heating season. Variable volume terminal reheat systems are sequenced if the cooling load is above minimum flow (usually 50%), and "reheat" below this point.

Reheat can be added in at least three different ways by central systems. The first is by direct operation of the zone thermostat, called "direct reheat." A 1-BTU reduction in zone cooling causes a 1-BTU increase in reheat.

The second type of reheat operates from outside temperature. Each drop of 1°F in outside temperature causes a specific increase in heat flow to the zone. The exact increase can be greater, smaller or equal to the change in zone load due to the 1°F change, so the zone thermostat may in turn respond by changing the zone cooling system output.

The third type of reheat involves mixing warm and cold air streams, as in most dual-duct and multi-zone type systems. The zone thermostat controls the proportions of air which flow to the zone from each of two ducts. In warm weather one duct is cold, the other is neutral. This mixing of cold and neutral air is not reheat. In cold weather, however, the "neutral" duct is heated to provide warm air, and the zone thermostat operates to mix warm and cold air. This results in some reheat, but not as much as by "direct reheat" control.

The system response of mixing systems is calculated by first solving this equation to find CFM air flow through each duct: $X\ (1.09)\ (RmT - Tc) = (\text{Zone CFM} - X)\ (Tw - RmT)\ (1.09) + \text{Zone SH}$

where X = Cold duct CFM
RmT = Room temperature
Tc = Cold duct temperature
Tw = Warm duct temperature

Cooling and heating coil loads are found by using the sensible heat formula:

$SH = 1.09\ (\text{coil CFM})\ (\text{Entering air temperature} - \text{Leaving air temperature})$

The action of the zone temperature controls may reduce air or water flow to the zone, cycle the zone fans, add reheat, or mix hot and cool air in order to keep zone capacity in balance with zone load. Zone response is frequently related to the percentage of maximum available cooling capacity. That is one reason why one needs to figure maximum design cooling capacity in step A1, and adjust this capacity for actual profile point operating conditions before calculating control response.

For example, assume a variable air volume – reheat type system with 50% maximum flow reduction and a zone installed cooling capacity of 26,000 BTUH sensible heat. If the zone cooling load were only 8,000 BTUH, zone control response would be to provide 50% of maximum capacity, or 13,000 BTUH cooling, and to balance this against the 8,000 BTUH room load by adding 5,000 BTUH of reheat.

Many systems have a minimum heating or cooling capacity. The perimeter office fan-coil system in the sample problem was provided with a constant supply of ventilation air from the interior office duct system, which was cold in summer and winter. The zone thermostat could not turn this cold air off, but could open the hot water valve if the room cooling load were less than this minimum cooling capacity of the ventilation air.

Zone system response calculations can be made on the same form as part of the zone thermal load calculations. If several types of systems are being compared, the completed thermal load portion of the form can be reproduced before calculating the system response.

Some types of central systems, such as VAV-reheat and VAV-dual duct, respond differently to different sizes of cooling loads. At 70% cooling load they use cooling energy only. At 30% cooling loads, they use cooling and some reheat, which also increases the cooling load. Reheat must be calculated before averaging zones or hours, for accuracy.

Assume a two-zone system with a VAV-reheat system, with 50% minimum air flow. Zone peak design loads might be:

	Zone peak loads	
Time	*East*	*West*
	10:00 AM July	4:00 PM July
BTUH	20,000	24,000
CFM at 20°F temperature difference (T.D.) (BTUH = (1.09 × T.D.)	917	1,101

The zone loads at 4:00 PM in July when the apparatus reached its peak load might be:

	Zone loads	
Time	*East*	*West*
	4:00 PM July	4:00 PM July
BTUH	7,000	24,000

Because the VAV – reheat system has a non-uniform response, one should estimate equipment load by using the zone-by-zone format. Assume the reheater is energized only when the zone load is less than 50% maximum.

50% VAV-reheat system	East	West
Actual zone thermal load @ 4:00 PM July	7,000	24,000
Minimum zone cooling capacity (50% of design load)	10,000	12,000
Reheat = Minimum capacity − Actual load	3,000	0 (cannot be negative)
Total cooling load = Thermal load + reheat	10,000	24,000
Zone CFM = Load + (1.09 × 20°F temperature difference)	459	1,101
Total CFM = 459 + 1,101 = 1560		

Had the block load method been used, the combined load of both zones would be more than 50% of the combined zone design load, so the reheater would be off, giving us:

Combined block load = 31,000 BTUH cooling, no reheat
Total CFM = 31,000 ÷ (1.09 × 20°F T.D.) = 1,422

The true block loads are 1,560 CFM, 34,000 BTUH cooling, and 3,000 BTUH reheat, so the figures were substantially low using the block load format.

Most systems have uniform system response at time of peak block load. At this time most zone loads are high enough to prevent this problem. Check to make sure before using the block load method.

To appreciate possible variations in zone response, compare the response of several types of systems to the same load. The assumption of zone thermal loads is:

Zone design thermal cooling load = 6,000 BTUH
Zone cooling design air flow = 6,000 ÷ (1.09 × 20°F) = 275 CFM
Actual zone heating load at 1:00 PM. November, 52°F outside temp. = 1,000 BTUH

VAV with 10% minimum air flow, sequenced heating control.

Min. VAV cooling = 10% × 6,000 (design) = 600 BTUH
VAV CFM = 10% × 275 (design) = 28 CFM
Heat must offset 600 BTUH of unnecessary cooling and handle heating load = 600 + 1,000 = 1,600 BTUH

VAV with 50% minimum air flow, sequenced heating control, supply air temperature reset to 10°F temperature difference.

Min. VAV cooling

$$= 50\% \times \frac{10°F\ T.D.\ actual}{20°F\ T.D.\ design} \times 6{,}000$$

= 1,500 BTUH
VAV CFM = 50% × 275 = 138 CFM
Heat must offset 1,500 BTUH of unnecessary cooling and handle heating load = 1,500 + 1,000 = 2,500 BTUH

Constant-volume dual duct with 85°F warm air, 55°F cold air, and 75°F room temperatures.

(Using formula, page 000): X (1.09) (75 − 55) = (275 − X) (85 − 75) (1.09) − 1,000
Cold duct flow = X = 61 CFM
Warm duct flow = 275 − X = 214 CFM

Conventional multi-zone with common heating coil serving all zones, mixing control, 85°F warm air, 55°F cold air, 75°F room temperatures.

Assume zone CFM increase due to reduced air friction with paralleled flow through both cooling and heating coils = 20%.
Actual zone CFM = 1.20 × 275 = 330.
Using mixing formula, X (1.09) (75 − 55) = (330 − X) 1.09 (85 − 75) − 1,000
Cold air flow = 79 CFM
Warm air flow = 251 CFM

Multi-zone with individual zone heating coils sequenced control, 55°F cold air, 70°F neutral air, 75°F room temperatures

Assume same zone CFM increase to 330, and minimum CFM (hot or cold damper leakage) = 10%.
Minimum zone cooling = 10% × 330 CFM × 1.09 × 20°F temp. difference = 719 BTUH
Minimum cold air flow = 10% × 330 CFM = 33 CFM
Neutral duct air flow = 330 − 33 cold air = 297 CFM
Heating load to warm neutral duct CFM to room temperature = 297 CFM × 1.09 × (75°F − 70°F) = 1.619 BTUH
Zone heating coil must offset minimum zone cooling (719) + warm-up neutral duct air (1,619) + handle zone heating load (1,000)
Heating coil load = 3,338 BTUH

2-pipe air and water induction system with 70 CFM of primary air supplied at 95°F from central apparatus, cold water in induction unit coil.

Heating effect of primary air = 70 CFM × 1.09 × (95 − 75) = 1,526 BTUH
Induction unit cooling coil must re-cool any excess heat supplied to room
Heat supplied (1,526) − heat load (1,000) = 526 BTUH cooling

Air entering the zone cooling or heating coil is sometimes taken directly from the room and is, therefore, at room temperature. Coil load can be calculated directly from zone loads, as with the 2-pipe induction system example above.

With the other systems, air entering the cooling or heating coils was probably not at room temperature. It was a mixture of return air from many different zones, and it had been subjected to heat gain or loss in the return duct system and then mixed with an unknown amount of outside ventilation air at 52°F. What can be determined from the zone response is how much air or water flow is needed, and what load this would represent if the coil entering air were at room conditions. It is usually more convenient to calculate zone response in BTUH than in CFM, but one must correct this load later as part of the central system response, to obtain the true coil loads.

Central system response is a term which includes:

- Calculating any item of the zone system response on a block load basis not calculated previously on a zone load basis. If the entire building thermal load was figured as a single block load, all zone and central system response calculations may be done here.
- Calculation of heat gain or loss in central system supply and return air duct, and pipe insulation losses.
- Calculating the effect of ventilation air and outside air economizer cycles on the central system coil loads. It is easy to figure the effect of ventilation air at the central system by calculating the outside and return air mixture temperature. Central cooling and heating coil loads are calculated from the air quantity and from entering and leaving air temperatures. This way of computing cooling coil loads is particularly useful if an outside air economizer cycle is being used, but some refrigeration is still required.
- Calculating the effect of "outside-air-to-exhaust-air" heat exchangers and "glycol run-around coil" systems for heat recovery. This is part of the system response process because it reduces the load presented to the heating and cooling equipment for the next step.
- Determining air and water flows required from central fans and pumps; also cooling and heating energy required from the central cooling and heating equipment.
- Determining the effect of the central fan arrangement on coil loads. Fan motor heat is added to the entering air temperature of a blow-through coil system and to the leaving air temperature of a draw-through coil system. Fan motor heat includes the heat equivalent of the entire fan motor power input if the motor is located in the conditioned space, in either a supply air or return air plenum. If the fan motor is outside the air stream in a non-conditioned equipment room, the useful work (about 90% of power input) is charged to the air stream, and motor inefficiency losses (about 10% of power input) are released into the non-conditioned equipment room.

The response of a central system to a number of zones with different loads can be complicated, and may differ for each type of system. Like the zone response, however, it is a vital step in preparing an accurate energy analysis.

Appendix B: Cost-Benefit Study

The cost-benefit study is usually the most important part of opportunity evaluation. The degree of detail depends on the client's needs and the extent of savings versus initial cost involved. Federal agencies may require life-cycle costing as a matter of policy, as do many local public agencies. When predicted savings far exceed all costs involved, a detailed life-cycle study is a waste of time. Private owners often pay only for opportunities which look feasible without a life-cycle study: that is, with less than a five-year simple payback. This is because

1 In the area of fuel energy, most costs are influenced by political decisions, and cannot be anticipated beyond the near term. In the private sector, people seldom invest where the degree of risk is not predictable.

2 Some owners may only require a simple cost-benefit statement. They prefer to keep financial information as to tax bracket, operations costs, costs of borrowing, etc., private. They also may want their own task force to make life-cycle studies, and they are in the best position to compare alternative investment plans.

In some cases, the costs of life-cycle analysis exceed the cost of the improvement opportunity being "costed." For example, moving an incorrectly placed thermostat could lead to big savings at little cost, and should be done without further ado. Methods have been developed in recent years in the field of "value engineering" and "engineering economics" to help make decisions regarding capital expenditures in buildings.

Simple methods for the economic evaluation of modifications include such basic concepts as savings/investment ratio, payback period, and BTU savings/investment dollars. These concepts are discussed in *"Life-Cycle Costing Emphasizing Energy Conservation Guidelines for Investment Analysis,"* a book published by the Energy Research and Development Administration (now part of the Department of Energy (Ref.), also ERDA-76/130).

Savings investment ratio

The savings/investment ratio (SIR) refers to present value of a modification's dollar savings divided by present value of its cost. Any item having a SIR greater than 1.0 may be considered cost-effective.

Converting savings and costs to their present value recognizes that a dollar today is worth more than a dollar in the future, because of the interest it can earn. Since the concept involves a consideration of time, a particular time period must be chosen as basis of the SIR. This is usually referred to as the useful life of the item. Also, a rate must be chosen to relate present dollars to future dollars. This is referred to as the discount rate, and can represent the interest rate or a desired rate of return on investment. A discount rate of 10% is common but the owner's requirements may call for use of a different discount rate. Tables with the useful life of various building types and pieces of equipment, along with tables containing factors for converting future savings and cost to their present value, appear in *"Life-Cycle Costing Emphasizing Energy Conservation Guidelines for Investment Analysis,"* (above) as well as in *"Life-Cycle Cost Analysis – A Guide for Architects"* (Ref.).

A factor called the differential cost escalation factor (DCE) may also be applied to take into account the relationship between predicted escalation rates and predicted inflation rates. A discussion of the DCE and accompanying tables may be found in the above ERDA publications.

Payback period

The most common method of cost/benefit analysis in the energy area is to look at the payback period. That is the time it takes savings resulting from an improvement to pay back its cost. The item's simple payback period is the number of years it takes for savings to equal initial cost. A discounted payback period takes into account the time value of money. In that case, the interest lost each year (due to the money used to pay for the item not being available for investment) is subtracted from the annual savings in calculating the payback period. This produces a longer payback period and is a more accurate measure of actual economic performance than the simple payback. It may, however, not be needed if the project results in a very short, simple payback period.

On many projects, owners will set a minimum payback period as an evaluation criterion. They also may have a required rate of return on investment. These figures should be established at the start of the project. Two items having the same SIR may have different payback periods. Thus, even though the SIR is often seen as a better index of economic performance, the payback period may be useful in choosing between alternatives. Cost effectiveness can be further improved if the owner can obtain a tax credit on the capital investment.

BTU savings/investment dollar

The third measure of cost-effectiveness of interest to an owner trying to save energy over time is the BTU savings/investment dollar ratio. This ratio is determed by dividing an improvement item's annual energy savings measured in BTU's by its annual present cost. Average annual present cost is the present cost of putting in the item divided by its economic life, which is the period over which it is expected to be the lowest cost alternative for accomplishing its purpose.

If a new product is involved or a better alternative is to come out in the near future, the item's economic life may be hard to determine. The useful life mentioned earlier can be used in most cases. The amount of confidence to place in the figure used for the item's economic life is the limiting factor in this method. An item's effectiveness in reducing peak energy demand cannot be evaluated by the BTU savings/investment dollar method.

All three methods are discussed in the above ERDA references.

Life-cycle cost study

If a more detailed cost-benefit analysis is called for, a life-cycle cost study should probably be undertaken. Life-cycle costing is an economic evaluation procedure developed by the armed forces to take into account all costs associated with each alternative over a specific time period, which is referred to as the "life-cycle." The modification's expected useful life is usually taken as its life cycle. When two alternatives with different useful lives are being considered, another figure may be used to provide a common basis of comparison. This method comes closest to considering all the criteria required for evaluation, but deals only with those which can be quantified in dollars.

A life-cycle cost analysis should take into account such costs as:

- Initial cost of carrying out the change, including design and possible interim financing.
- Cost of money or interest.
- Annual operation costs such as fuel, utilities, services, operation personnel and normal maintenance.
- Anticipated repairs and parts replacements.
- Total replacement cost at the end of the item's useful life, including disposal costs.
- Income tax consequences.
- Any additional costs or credits possibly associated with the item, such as increased or reduced wear on other parts of the building or on pieces of equipment not part of the item.
- All revenue charges caused by the item, such as lowered income resulting from reduced hours.
- Credit for salvage or residual value if expected useful life of the item exceeds the life-cycle being considered.

The owner's specific requirements will determine which of these costs to consider in each project and how accurately to assess them.

This technique is especially useful when comparing an alternative having a high initial cost with good efficiency and low maintenance to one which may cost less at first but have poor efficiency or high maintenance costs. The choice applies most to new buildings where initial cost is high. It is also useful when comparing an item's expected savings to total costs involved whenever there is doubt as to its effectiveness.

In a life-cycle cost analysis, associated costs and credits are either converted to their equivalent present value or to an equivalent uniform annual value, for comparison, using appropriate conversion factors. A summary of these factors is given in *"Life-Cycle Cost Analysis – A Guide for Architects"* along with a good bibliography. There is also a discussion of how to apply the refinements of life-cycle costing to the simple economic evaluation techniques.

One limitation of life-cycle analysis is the difficulty of placing dollar figures on improved comfort, better appearance, and the impact of change and pleasant surroundings on the work force.

Sensitivity analysis
Sensitivity analysis is a further refinement of economic evaluations. It is a method for determining the effect on an original analysis if a variable is changed. Sensitivity of the original analysis to a change in input variables is determined. Detailed evaluation of an energy planning project necessarily involves sensitivity analysis. That is because the main evaluation technique is to determine the change in the building's performance resulting from carrying out an improvement, and this is done by changing input variables.

Sensitivity analysis is particularly useful when the item being considered appears marginal. For example, if the projected rate of increase in costs, or the rate of inflation used in the original economic analysis, is open to question, a sensitivity analysis could determine if the earlier conclusions are still valid in light of these changes. In essence, sensitivity analysis amounts to redoing the original analysis with new input and comparing the results.

Economic leverage
Economic leverage is another factor when evaluating opportunities.

If an existing building can save one dollar per square foot per year in energy costs and this dollar goes into increased profit, then the building can command an additional sales price of $10 per sq. ft. assuming a capitalization rate of 0.10. Therefore, a 100,000 sq. ft. building would sell for $1,000,000 more and the owner could afford a thorough energy planning project.

Glossary

This glossary includes terms used in this book as well as a few other terms frequently used in energy analysis work.

Air Conditioning
The process of treating air so as to control simultaneously its temperature, humidity, cleanliness and distribution to meet the comfort requirements of the building's occupants.

ANSI
American National Standard Institute, Inc.

ASHRAE
American Society of Heating, Refrigerating and Air Conditioning Engineers.

ASHRAE Standard 90-75
Energy Conservation "component performance" standards for new building design developed and adopted by ASHRAE in 1975.

Billing Demand
The demand figure used to determine the monthly demand charge. It may be different than the actual month's peak demand if the utility rate structure includes a clause whereby the highest recorded demand from a previous month may govern the monthly demand charge.

Billing Tape Recorder (BTR)
A demand meter which records energy use on a magnetic tape cassette. It is more accurate and easier for the utility company to read but does not display information as conveniently as a graphic recording meter.

Bin
An increment used to simplify a range of data. For instance if the outside temperatures are grouped into increments of 5 degrees, each increment is referred to as a bin.

Bin Method
A method for estimating the energy use of equipment which responds to changes in weather, primarily HVAC equipment, utilizing weather data that has been organized into temperature bins. (5° is the usual increment.)

Block Load
A thermal load calculated by combining several different thermal zones rather than calculating the load for each zone separately. The load on a group (or "block") of zones, usually calculated by combining the areas and other load components so that the calculations are made as if the many small zones were really one large zone.

Boiler Capacity
The rate of heat output in BTUH (W) measured at the boiler outlet at the design inlet, outlet, and rated input conditions.

British Thermal Unit (BTU)
A unit of measurement equivalent to the amount of heat energy required to raise the temperature of one pound of water one degree Fahrenheit.

BTUH
An abbreviation for BTU per hour.

Building Envelope
The elements of a building (e.g., walls, roofs, floors) enclosing conditioned spaces through which thermal energy may be transferred.

Bypass
A pipe or duct, usually controlled by valve or damper, for conveying a fluid or gas around an element of a system.

Capacity, Thermal or Heat Storing
The property of a material which determines the amount of heat or cooling it will absorb. It is a function of the mass of the material and its specific heat.

Climatological Design Conditions
Selected outdoor conditions which predict the maximum and minimum temperatures to which a building will be subjected. Also, outdoor temperature conditions selected to maintain occupants' comfort.

Coefficient of Heat Transmission (U-value)
The time rate of heat flow through a body per unit area for a unit temperature difference between the fluids on the two sides of the body under steady-state conditions.

Coefficient of Performance
The ratio of heating or refrigeration system effect to the rate of energy input, in consistent units, under designated operating conditions.

Cogeneration
The simultaneous production and use of heat and electricity.

Coincident Wet Bulb-Temperature
The wet bulb temperature reading that is taken at the same time as a dry bulb temperature reading is called a "coincident wet bulb temperature" because it is related to a specific dry bulb temperature.

Comfort Envelope
The area on a psychrometric chart enclosing all those conditions described in ASHRAE Standard 55-74 "Thermal Environmental Conditions for Human Occupancy," as being comfortable.

Comfort Zone (average)
The range of effective temperatures over which the majority (50 percent or more) of adults feels comfortable; (*extreme*) the range of effective temperatures over which one or more adults feel comfortable (see Effective Temperature).

Conductance, Thermal
(c) The time of heat flow through a body per unit area for a unit temperature difference between the body's surfaces under steady-state conditions. Usually stated for a specific thickness rather than a unit thickness.

Conduction
The method whereby heat is transferred from one body to another without physical displacement of the matter within the bodies.

Conductivity, Thermal
(k) The time rate of heat flow through a homogeneous material per unit area and thickness, under steady-state conditions, when a unit temperature gradient is maintained in the direction normal to the cross-sectional area.

Control, Automatic
A device used to regulate a system on the basis of the response to changes in the magnitude of some property of the system, e.g., pressure, temperature.

Control Point
A control setting at which the operation of the equipment is changed, such as a temperature at which it shuts off. Also called a control point setting.

Consumption (energy)
The total amount of energy used over a specific period of time, usually a month or a year. The common unit of energy consumption for buildings is the BTU. Each energy source will have its own unit of consumption for billing, which can be converted to BTU's. For electricity it is the kilowatt-hour, for gas it is usually the therm or a multiple of cubic feet, and for most fuels it is the quantity of fuel such as the gallon or pound.

Convection
The transfer of heat by movement of a fluid (gas, vapor, or liquid).

Convection, Forced
Convection resulting from circulation of a fluid by a fan or pump.

Convection, Natural
Circulation of a gas, vapor or liquid medium (usually air or water) due to differences in density resulting from temperature changes.

Cooling Load
The rate at which heat must be removed by the cooling equipment to maintain indoor comfort conditions. (The cooling load differs from heat gain in that the radiant part does not immediately appear as the cooling load but is absorbed by surfaces that enclose the space. When these surfaces become warmer than the indoor air, heat is transferred to the air by convection.)

Daylighting
The use of controlled natural lighting methods indoors through toplighting (skylights), sidelighting (windows), and/or uplighting (reflection).

Degree Day, Heating
A unit measurement based on temperature difference and time, used in estimating average heating requirements for a building. For any one day, when the mean outside temperature is less than 65°F, there exist as many degree days as there are Fahrenheit degrees difference in temperature between the mean temperature and the base temperature 65°F. This base temperature assumes that no heat input is required to maintain the inside temperature at 70°F when the outside temperature is 65°F.

Dehumidification
The condensation of water vapor from air by cooling below the dewpoint or removal of water vapor from air by chemical or physical methods.

Demand
The rate at which energy or fuel is consumed by a piece of equipment, or by the building as a whole. Demand is most often measured for electricity and is usually expressed in kilowatts. Most electric companies base their demand charges on the highest amount of energy consumed per 15 or 30 minutes each month. This is an "integrated" demand. Some companies may use the highest instantaneous rate of energy consumption instead, or an instantaneous demand.

Demand Charge
A charge, based on demand, included in utility bills.

Demand Limiting (load leveling)
An equipment control method which aims at reducing peak energy demands by turning equipment off usually to defer loads as a peak is approached according to an established priority schedule. Manual Demand Limiting relies on building operators to predict peak periods based on operation records and experience. Automatic Demand Limiting Equipment monitors energy use and shuts equipment down as a peak is approached.

Demand Meter
A meter which measures demand. Usually a recording electric meter which records the maximum demand for a period of time (usually ¼ or ½ hours) and then resets to 0.

Design Outside Temperature
The outdoor temperature used in the calculation of the heating load. This temperature is derived by statistical methods and is not necessarily the lowest temperature ever recorded for a given locality.

Diurnal Temperature Range
The range of temperature occurring over a 24-hour time span.

Diversity Factor
A factor applied to heating or cooling loads to account for the likely non-occurrence of part of the load. The diversity factor used will be based on judgment and will vary with the load and conditions being considered.

DOE:
Department of Energy of the United States

Draw
Often used to refer to the instantaneous rate at which a piece of equipment consumes energy, its demand.

Dry Bulb Temperature
The measure of the sensible temperature of air.

Dual Duct System
HVAC system that uses two separate ducts to each outlet. One duct supplies hot air and the other supplies cold air. Mixing the two airstreams at the outlet provides the desired temperature.

Duty Cycle
An equipment control method which aims at using equipment only when it is needed. The term is most often applied when timers are used to turn equipment on and off.

Economizer Cycle
HVAC system that uses cool, outdoor air, whenever possible, to offset heat gains in the building rather than using a refrigeration machine, which requires electrical energy to drive it.

Edge Loss
The heat loss where the exterior walls of a building meet the ground floor slab.

Effective Temperature
An arbitrary index which combines into a single value the effect of temperature, humidity and air movement on the sensation of warmth or cold felt by the human body.

Efficiency, Thermal
Relating to heat, a percentage indicating the available BTU input converted to useful purposes. The term is generally applied to combustion equipment (E = BTU output/BTU input).

Electric Demand Limiter
A device that turns off or reduces the power to electrically powered equipment whenever that total electrical load rises to a predetermined level.

Emissivity, Thermal
The capacity of a material to emit radiant energy. Emittance is the ratio of the total radiant energy emitted by a body to that emitted by a "black body," (a perfect absorber of radiant heat that emits none) at the same temperature.

Energy
The capacity for doing work. Different forms may be transformed from one type into another, such as thermal (heat), mechanical (work), electrical, and chemical. Energy is measured conventionally in kilowatt-hours (KWH) or British Thermal Units (BTU). Energy is measured in SI units in joules (J) where 1 joule = 1 watt second.

Energy Analysis
A comprehensive study of a building's energy use and costs to determine how much, when, why, and where energy is used in the building. It will allow the energy planner to determine the feasibility of the opportunities identified by the energy audit as requiring further study and to identify and evaluate many new opportunities as well. It may also be used to determine energy code compliance since it includes a mathematical simulation of the building's energy use.

Energy Audit
A specific accounting of the various forms of energy used during a designated period of time, usually annually. It aims at identifying the major opportunities for improving a building's energy performance through a short study of the building; its operation and use; its equipment; and its energy bills. It can also be used to determine energy code compliance in some cases.

Energy Budget
A design or consumption budget for a building or space use which is expressed in energy units that reflect the energy used to operate the various service systems. (Budget figures may also take into account the nature of the energy source, including conversion, transportation and distribution losses.)

Energy Management System
An automatic building control system which may include demand limiting, duty cycling, equipment startup and shutdown, and building monitoring and utilizes a strategy designed to maximize energy efficiency and minimize energy costs.

Energy Profile
The relationship between energy consumption or demand and outside air temperature, usually plotted at appropriate "temperature bins."

Enthalpy
Thermodynamic property of a substance defined as the sum of its internal energy plus the quantity Pv/J, where P = pressure of the substance, v = volume of substance, and J = the mechanical equivalent of heat.

Enthalpy Control
An HVAC control system which compares the temperature and the humidity of the outside air against that of the return air, determines which will result in the lowest cooling load, and adjusts the mixed air dampers accordingly. Used in combination with an outside air economizer.

Equipment Performance
The input requirements and useful output of equipment under a specific operating condition. Most HVAC equipment operates under widely varying conditions with significant changes resulting in input and output. Input and output can include source energy, as well as cooling and heating energy, air and water flow, temperatures, pressure drops.

Equivalent Sphere Illumination (ESI)
A standard of illumination measurement used to evaluate quality of illumination. The level of illumination required from a uniformly luminous sphere around the task (similar to conditions of complete cloud cover) to achieve the same task visibility as the actual situation.

ERDA
Energy Research and Development Administration. (Now part of DOE).

Equivalent Full-Load Hours
Total energy consumption divided by the Full-Load energy input. This gives the number of hours a piece of equipment would need to operate at its full capacity to consume as much energy as it did operating at various part-loads.

Exfiltration
Indoor air leakage to the exterior through building envelope caused by a pressure differential.

Fahrenheit
Thermometric scale in which 32 degrees denotes freezing and 212 degrees denotes the boiling point of water under normal pressure at sea level (14.696 psi).

Fan Coil
A heating and/or cooling device that blows air through a coil.

FEA
Federal Energy Administration.
(Now part of DOE.)

Fin Tube
Tube with metal sheets attached to increase heat transfer area.

Footcandle
A unit of measurement for the average illumination on a surface. One footcandle equals one lumen per square foot.

Forced Air Heating System
A heating system in which air is circulated mechanically, utilizing either a blower or fan as the transfer medium.

Fossil Fuels
Decayed organic matter stored within the earth, transformed over millions of years into coal, petroleum, natural gas, and peat.

Fuel Adjustment
A charge used by utility companies to pass on to the customer the difference between the actual fuel cost to the utility and the fuel cost the utility has included in its basic rate.

G-9 Chart
A circular chart produced by a General Electric G-9 recording demand meter which graphically records the peak demand during each demand period for 32 days.

Geothermal Energy
Large underground reservoirs of steam and scalding water found from a few hundred to 30,000 feet beneath the surface.

Gigawatt
1,000 megawatts or one million kilowatts.

HVAC
Heating, ventilating, and air conditioning.

HVAC System
A system that provides heating, ventilating, and/or air conditioning within or associated with a building.

Heat
The form of energy that is transferred by virtue of a temperature difference.

Heat Capacity
The quantity of heat required to raise the temperature of a given mass of a substance one degree.

Heat Exchanger
A device used to transfer heat from a fluid flowing on one side of a barrier to a fluid flowing on the other side of the barrier. Quite often this is done by running a coil of pipe through a tank.

Heat Gain
As applied to HVAC calculations, it is that amount of heat gained by a space from all sources, including people, lights, machines, sunshine, etc. The total heat gain represents the amount of heat sensible and latent, that must be removed from a space to maintain desired indoor conditions. (See Cooling Load.)

Heat Island
The occurrence of generally higher ambient air temperature in urban areas due to high energy consumption, heat buildup, and large amounts of surface area that absorbs heat.

Heat Lag
Time delay of heat transfer through material due to heat capacity and thermal resistance.

Heat Loss
The sum cooling effect of the building structure when the outdoor temperature is lower than the desired indoor temperature. It represents the amount of heat that must be provided to a space to maintain indoor comfort conditions.

Heat Pump
A refrigeration system designed so that the heat extracted at a low temperature and the heat rejected at a higher temperature may be used alternately for heating and cooling functions respectively. (See Unitary Heat Pump.)

Heat Recovery
Heat utilized which would otherwise be wasted.

Heat Sink
A body (water, earth, metal, etc.) capable of accepting and storing heat. It can also serve as a heat source.

Heat Transfer
The methods by which heat may be propagated or conveyed from one place to another. This may be by conduction, convection or radiation.

Heating Load
The rate at which heat must be provided by the heating equipment to maintain indoor comfort conditions. This is seldom the same as the heat loss, since the building may have internal sources of heat gain which counteract some of the heat loss.

Horsepower
A standard unit of power equal to 746 watts in the United States. One horsepower equals 2,545 BTU (mean) per hour, 550 foot-pounds per second.

Humidity, Relative
The ratio of the amount of water vapor actually present in the air to the greatest amount possible at the same temperature.

Infiltration
Outdoor air leakage into a building. It most often occurs at cracks around doors, windows and other openings and is caused by the pressure effects of wind and the effect of differences in the indoor and outdoor air density.

Insolation
The solar radiation incident at the earth's surface.

Insulation
A material having a relatively high resistance to heat flow and used principally to retard the flow of heat. Four major classifications of building insulating materials are: 1) batt, 2) loose fill, 3) reflective, and 4) rigid.

Integrated Demand
The total energy consumed in a time period divided by the time expressed in hours. Note that this can be considerably different than instantaneous demand. Usually a ½ hour period is used.

Kilowatt (kw)
A unit of power equal to 1,000 watts, or to energy consumption at a rate of 1,000 joules per second. It is usually used for electrical power.

Langley
The meteorologist's unit of solar radiation intensity, equivalent to 1.0 gram calorie per square centimeter. (1 Langley per minute = 221.2 BTU per sq. ft.)

Latent Heat
The change in heat quantity that occurs without any corresponding change in temperature. Usually accompanied by a change of state; for example, water may change to steam, or the moisture content in air may be increased.

Life-Cycle Cost Analysis
The process of accounting for the capital and operational costs over the useful life of a building.

Load
Indicates a rate of flow of energy for either a heating or cooling requirement or a total of both (expressed in terms of BTU per hour, BTU per month or BTU per year).

Load Factor
The total monthly consumption divided by the monthly demand and then divided by the total number of hours in the month. It is less than one and can be expressed as a percentage.

Load-Shedding Device
A mechanical contrivance that will enable the load of a machine to be reduced on demand.

Lumen
Measure of light intensity; one lumen per square foot equals one footcandle.

Luminaire
A complete lighting unit consisting of a lamp or lamps together with the parts designed to distribute the light, to position and protect the lamps and to connect the lamps to the power supply.

Mean Radiant Temperature (MRT)
A weighted average of the various radiant influences in a space taking into account surface temperatures and surface exposure angles (relative to a person's location) in degrees. If a person is assumed to be a cylinder and radiation effects of floor and ceiling are ignored MRT equals the sum of the arcs multiplied by the surface temperatures of each individual arc.

Megawatt
A unit of power equal to 1,000 kilowatts or 1 million watts.

Methane
A colorless, odorless, flammable gaseous hydrocarbon (CH_4) which is the product of the decomposition of organic matter. It is the major component of natural gas.

Microclimate
Climate at specific site as defined by local variations in the regional climate caused by topography, vegetation, soils, water conditions, as well as man-made construction.

Multi-Zone System
A HVAC control air-handling unit that provides hot and cold air and facilities for mixing these at the unit to give different supply temperatures to a number of zones.

Nameplate Rating
A statement by the manufacturer on a mechanical device that gives the performance including energy draw of the system under specific operating conditions.

Night Set-Back
An HVAC control strategy which maintains lower room temperatures during unoccupied periods in the heating season.

Night Set-Up
An HVAC control strategy which maintains higher room temperatures during unoccupied periods in the cooling season.

NOAA
National Oceanic and Atmospheric Administration.

Non-Renewable Energy
A depleting energy source which is virtually impossible to recreate, such as the fossil fuels (oil, coal, natural gas).

"Off-Peak" Generation
The practice of generating hot or cold water in a building during the hours when utility rates are lowest and storing it for later use. Especially advantageous when demand charges are high or when "time-of-day" rates are employed.

Outside Air
Air taken from the outdoors and, therefore, not previously circulated through the system.

Passive Design Methods
Using building elements to accept, store, and distribute the flows of energy from nature (most often solar) to provide benefits to activities and processes within the building without the use of external power. If external power is required for distribution by fans or pumps it is called a "hybrid" system.

Payback Period
The amount of time it takes the savings resulting from a modification to "pay back" the costs involved. A "simple" payback period does not consider the time value of money. A discounted payback period does.

Peak Energy Demand
The maximum demand recorded during the metering period, usually a month.

Peak Load
Maximum predicted load over a given segment of time for any system.

Percent Sunshine
The amount of sunshine on an average day each month given as a percentage of the amount of sunshine on a perfectly clear day for that month.

Performance Standards
Standards describing performance in terms of minimum energy required of entire buildings and their energy using systems based on weather conditions, occupancy patterns, type of energy source and other pertinent energy use factors.

Plenum
A compartment for the passage and distribution of air often above a ceiling.

Power
In connection with machines, power is the time rate of doing work. In connection with the transmission of energy of all types, power refers to the rate at which energy is transmitted; in customary units, it is measured in watts (W) or British Thermal Units per hour (BTUH); in SI units it is measured in watts.

Power Factor
The ratio between actual electric power consumption in watts and the theoretical power obtained by multiplying volts by amperes. When the power factor is unity, $kv \times a = kw$. This is the ideal situation and can be achieved within practical limits, by installing capacitors on inductive circuits.

Prescriptive Standards
Design standards stipulating procedures in terms of specific materials and components rather than building performance.

Process Energy
The energy consumed by any processing or manufacturing equipment (or energy used for purposes other than environmental needs).

Profile
See Energy Profile.

Profile Point
Energy consumption or demand plotted at a specific outside temperature used to create the "Energy Profile."

Psychrometric Chart
A standard chart showing the relationship between the dry-bulb temperature, the wet-bulb temperature, and the moisture content of air, often used for estimating air conditioning loads.

Psychrometric Process
Any process which involves changing the thermal properties of moist air. A process that can be plotted on a psychrometric chart.

Psychrometrics
A science which studies the thermodynamic properties of moist air and the effect of atmospheric moisture on materials and human comfort. Also used to describe the method of controlling the thermal properties of moist air.

Quad
10^{15} BTUs (one quadrillion BTU).

R-Value
Thermal resistance equal to (See U-value).

Radiation
Energy in the form of electromagnetic waves which is continually passing between the surfaces of all bodies.

Radiation, Solar
Radiant energy emitted from the sun in the wavelength range between 0.3 and 3.0 microns. Of the total solar radiation reaching the earth, approximately 3% is in the ultraviolet region, 44% in the visible region, 54% in the infrared region. Radiant energy may be expressed as 1) diffuse – solar radiation received from the sun after its direction has been changed by reflection and scattering by the atmosphere, or as 2) direct beam – solar radiation received from the sun without undergoing a change of direction.

Radiation, Visible (light)
Radiant energy of wavelengths from 0.4 to 0.76 microns which produces a sensation defined as "seeing" when it strikes the retina of the human eye.

Ratchet Clause
A clause in a utility rate structure by which the monthly demand charge may be governed by the highest recorded demand (ratchet) from a previous month.

Recovered Energy
Energy utilized which would otherwise be wasted.

Reflectivity
The capacity of a material to reflect radiant energy. Reflectance is the ratio of the radiant energy reflected from a body to that incident on it.

Refrigerant
The fluid used for heat transfer in a refrigerating system. It absorbs heat at a lower temperature and a low pressure of the fluid and rejects heat at a higher temperature and a higher pressure of the fluid. Usually changes of state of the fluid are involved.

Reheat
The application of sensible heat to supply air that has been previously cooled below the temperature of the conditioned space by either mechanical refrigeration or the introduction of outside air to provide cooling.

Renewable Energy
Sources of energy derived from incoming solar radiation (including photosynthetic processes) from phenomena resulting therefrom (including wind, waves and tides, lake or pond thermal differences) and energy derived from the internal heat of the earth (including nocturnal thermal exchanges). (Synonyms include non-depletable, current income energy, regenerative energy.)

Reset
Adjustment of the set point of a control instrument to a higher or lower value automatically or manually to conserve energy.

Resistance, Thermal (R value)
The reciprocal of thermal conductance.

Return Air Load
The portion of the sensible load due to heat gained or lost directly in the return air plenum.

Sensible Heat
Heat that results in a change in temperature (as opposed to latent heat) and that can be "sensed," or felt.

Shading Coefficient
The ratio of the solar heat gain through a glazing system corrected for external and internal shading to the solar gain through an unshaded single light of double strength sheet glass under the same set of conditions.

SI Units
International system of units (metric) from the French "Le System International d'units."

Solar Energy
Energy derived from the sun in the form of electromagnetic radiation.

Solar Heat Gain
The amount of sensible heat gained by a space from the sun equal to the solar energy striking the glass reduced by a shade factor to compensate for the solar energy transmittance characteristics of the type of glazing used and the effectiveness of any shading devices.

Specific Heat
The amount of heat that has to be added to or taken from a unit of weight of a material to produce a change of one degree in its temperature.

Steady State
Refers to steady state temperature conditions which are often assumed when estimating heat flow through the building envelope. This ignores the effects of changing outdoor temperatures and thermal capacity. It is generally regarded as an accurate assumption when average loads are being calculated particularly if the temperature difference between indoor and outdoor air is great with respect to the short-term changes in the outdoor air temperature or if the heat storing capacity of the building is small with respect to the total heat flow as is the case when lightweight building materials are used. It is not considered accurate for determining peak loads especially in buildings with high heat storing capacity or in climates with large daily variations in outdoor air temperature and solar radiation.

Storage, Heat
The amount of heat that can be stored by a substance.

Sun Time
A time related directly and exclusively to the position of the sun. For example, noon occurs when the sun is due south. There are variations in different longitudes.

Surge
A high rate of energy use over a short period of time usually used with reference to the "starting surge" necessary to start-up electric motors. This surge will have a significant effect on "instantaneous" demand but only a small effect on "integrated" demand.

System
A combination of equipment and/or controls, accessories, ductwork and piping, and terminal devices arranged to perform a specific function, such as HVAC, service water heating or illumination.

System Performance
The input and output of the complete system at a specific operating condition. Equipment performance is an important determinant of system performance.

System Response (HVAC)
The response of a particular HVAC system to a specific or a particular building thermal load. System response varies with system arrangement and control and generates the output requirements of the HVAC equipment.

Temperature
A measure of heat intensity or the ability of a body to transmit heat to a cooler body.

Terminal Reheat System
A HVAC system that cools air down to a constant condition regardless of load, or to a condition that will satisfy the zone having the maximum requirement, and then reheats it to meet building requirements.

Therm
A quantity of heat equal to 100,000 BTUs.

Thermal Zone
One or more spaces within a building that have similar cooling and heating load patterns and can therefore have common HVAC system controls.

Thermostat
An instrument which measures changes in temperature and controls devices for maintaining a desired temperature.

Time Lag
The delay caused by heat storage in a building element (such as a wall) and its subsequent release by the structure. As the mass of the element increases, the time lag increases.

"Time-of-day" Rates
Rate structures used by utility companies which charge higher rates during the high use periods of the day to discourage energy use during high use hours, reduce peak generating demands, and encourage more even and efficient use of electricity generating plants.

Ton of Refrigeration
One ton of refrigeration means the removal of heat at the rate of 12,000 BTUs per hour. This unit comes from the fact that melting a ton (2,000 lbs.) of ice requires 2,000 × 144 BTU, which if done over a 24-hour period, requires a heat removal rate of 12,000 BTU per hour.

Total Energy System
An on-site electrical plant with the capacity to recover and reuse its own waste heat.

Transmission, Thermal
The rate at which heat passes through a material; directly proportional to the U-value and the temperature differential across the material.

Transmissivity
The capacity of a material to transmit radiant energy.

Transmittance (of solar energy and light)
The ratio of radiant energy transmitted through a material to energy incident on a surface.

TRY
ASHRAE Test and Reference Year data. The future standard weather data for the air conditioning and refrigeration industry.

U-Value
The heat flow rate through a given construction assembly, air to air, expressed in BTU/Hr./Sq. Ft./ degree difference between indoor and outdoor temperatures. The coefficient of heat transmission for a construction assembly.

Uniform Response Equipment
Equipment that operates with constant energy input whenever it is operating regardless of conditions.

Unitary Cooling and Heating Equipment
One or more factory-made assemblies which normally include an evaporator or cooling coil, a compressor and condenser combination, and may include a heating function as well. Where such equipment is provided in more than one assembly, the separate assemblies shall be designed to be used together.

Unitary Heat Pump
One or more factory-made assemblies which normally include an indoor conditioning coil, compressor(s) and outdoor coil or refrigerant-to-water heat exchanger, including means to provide both heating and cooling functions. It is designed to provide the functions of air-circulating, air cleaning, cooling and heating with controlled temperature, and dehumidifying, and may optionally include the function of humidifying. When such equipment is provided in more than one assembly, the separate assemblies shall be designed to be used together.

Vapor Barrier
Any thin, waterproof membrane used to substantially reduce passage of moisture, such as under concrete slabs, in wall assemblies or roofs.

Variable Volume System
A HVAC system that operates at a constant temperature and varies the volume of air supplied to any space in accordance with its cooling requirements.

Ventilation
The process of supplying or removing air, by natural or mechanical means, to or from any space. Such air may or may not have been conditioned.

Ventilation Air
That portion of supply air which comes from outside (outdoors) plus any recirculated air that has been treated to maintain the desired quality of air within a designated space.

Watt
The amount of work available from an electric current of 1 ampere at a potential of 1 volt. The watt is also the metric unit of power, and is equal to a rate of energy consumption of 1 joule per second. One joule is roughly one thousandth of a BTU; 1,000 watts (1 Kilowatt), therefore, is roughly the amount of energy required to raise the temperature of one pound of water one degree Fahrenheit per second.

Wet-Bulb Temperature
The lowest temperature attainable by evaporating water in the air without the addition or subtraction of energy (adiabatic saturation).

Zone
See Thermal Zone.

References

American Institute of Architects. **"AIA Energy Notebook."** (Washington, D.C.: The American Institute of Architects, 1735 New York Avenue, N.W., Washington, D.C., 20006, 1975).

An information service on energy and the built environment including opportunities, case studies, articles, tools/techniques, references, quarterly updates and monthly newsletters.

_______________. **Energy and the Built Environment: A Gap in Current Strategies.** (Washington, D.C.: A.I.A., May, 1974).

First energy report by AIA that defines energy efficient buildings; potential of energy savings offered by the building sector; and evolves an approach for achieving that potential. Also establishes a long term action plan for AIA energy programs.

_______________. **Life Cycle Cost Analysis, A Guide for Architects.** (Washington, D.C.: A.I.A., 1977).

_______________. **A Nation of Energy Efficient Buildings by 1990.** (Washington, D.C.: A.I.A., February, 1975).

Shows the financial feasibility and an implementation strategy for capturing the full potential of energy savings quantified in Energy and the Built Environment: A Gap in Current Strategies.

_______________. "Regional Climate Analysis Design Data: The **House Beautiful** Climate Control Project." **Bulletin of the American Institute of Architects.** September, 1949, through January, 1952. (Available from Xerox University Microfilms, P.O. Box 1467, Ann Arbor, Michigan, 48106).

Climatological and design data for 15 population nodes in the U.S. Weather data on temperature, sun, wind, and precipitation. Oriented toward residential construction.

ASHRAE Handbook and Product Directory: 1974 Applications. (New York: American Society of Heating, Refrigerating and Air-Conditioning Engineers, Inc.; 345 East 47th Street, New York, New York 10017).

Technical information and data on HVAC components, units and systems is presented.

_______________: **1975 Equipment.** (New York: American Society of Heating, Refrigerating and Air-Conditioning Engineers, Inc.).

Information on available types and capacities, principles of operation, construction, performance, testing and rating, and a selection of considerations for various HVAC components and assemblies.

_______________. **1976 Systems.** (New York: American Society of Heating, Refrigerating and Air-Conditioning Engineers, Inc.).

Information on available types, performance characteristics, application criteria, design considerations and component selection for various HVAC systems.

_______________. **1977 Fundamentals.** (New York: American Society of Heating, Refrigerating and Air-Conditioning Engineers, Inc.).

Covers theory and basic data essential for HVAC equipment development, system design and application.

ASHRAE Cooling and Heating Load Calculation Manual (1979).

Provides data and a method of calculating cooling and heating loads.

ASHRAE Standard 90-75: Energy Conservation in New Building Design. (1975).

Component performance requirements and criteria directed toward the design of building envelopes with high thermal resistance, low air leakage, and toward improved design of mechanical and electrical systems.

ASHRAE Standard 62-73: Standards for Natural and Mechanical Ventilation. (1973).

ASHRAE Standard 90-75 Section 12: Annual Fuel and Energy Resource Determination. (1975).

Addendum to ASHRAE Standard 90-75 which provides for a method of reporting the calculated quantities of fuel and energy resources anticipated to be consumed in meeting the annual needs of a proposed building project.

Carrier Air Conditioning Company. **Carrier System Design Manual, Part 1, Load Estimating.** (New York, New York: McGraw-Hill, 1972).

A concise design guide for sizing air conditioning systems.

Caudill, William W., et. al. **A Bucket of Oil: The Humanistic Approach to Building Design for Energy Conservation.** (Boston: Cahners Books, 1974).

Identifies opportunities and suggests ways for saving energy through building design.

Dubin, Fred S. and Chalmers G. Long, Jr. **Energy Conservation Standards for Building Design, Construction, and Operation.** (New York: McGraw-Hill, 1978).

Coxe, Weld. **Marketing Architectural and Engineering Services.** (New York: Van Nostrand, 1971).

Practical tools and techniques are offered for the process of new business development. Concepts are applicable to entering new energy related fields.

Crowther, Richard. **Sun/Earth: How to Apply Free Energy Sources to Our Homes and Buildings.** (Denver: Crowther/Solar Group, 1976). (AIA Publications Marketing).

Nontechnical overview and introduction to fundamental energy concepts, pros and cons of fossil fuels, and discussion of regenerative energy forces. Sample projects by author display the author's own consciousness about the energy-architecture relationship.

Educational Facilities Laboratories, Inc. **Energy Conservation and the Building Shell.** (Menlo Park: EFL, Inc., 1974).

An excellent, short, realistic description of the usual energy-related problems in schools.

Egan, M. David. **Concepts in Thermal Comfort.** (Englewood Cliffs, N.J.: Prentice-Hall, 1975).

Theory, data and calculation methods for building thermal design, climatic considerations, and HVAC systems are presented in a graphic format.

Fanger, P.O. **Thermal Comfort: Analysis and Applications in Environmental Engineering.** (New York: McGraw-Hill, 1973).

Develops rational basis for the establishment of thermal comfort conditions, for the evaluation of thermal environments, and for carrying out environmental analysis.

Federal Energy Administration. **Guidelines for Saving Energy in Existing Buildings.** (Washington, D.C.: Government Printing Office, 1975-1976, two volumes).

Opportunities and methods for reducing energy consumption in existing buildings. Volume 1 is directed toward building owners and operators; Volume 2 is for architects, engineers, and others responsible for analyzing and devising more comprehensive approaches.

Franta, Gregory E. and Kenneth R. Olson. **Solar Architecture.** (Ann Arbor: Science Publishers Inc., 1978).

Suggests new approaches to architecture using case studies incorporating life style changes, solar heating, and food production.

General Services Adminstration. **Energy Conservation Design Guidelines for New Office Buildings,** 2nd edition. (Washington, D.C.: Government Printing Office, 1975).

Checklist of energy saving opportunities with discussions of energy design concepts. Establishes a total building energy budget of 55,000 BTU/GSF/YR.

____________. **Energy Conservation for Existing Office Buildings.** (Washington, D.C.: Government Printing Office, 1975).

Companion publication of Energy Saving Conservation Design Guidelines for New Office Buildings, *2nd ed. Establishes goal of 75,000 BTU/GSF/YR for building energy use. Contains both design and operation guidelines.*

Givoni, B. **Man, Climate and Architecture,** 2nd edition. (London: Applied Science Publishers, Ltd., 1976).

Discusses the physiological, physical and architectural aspects of the relationships between climate, man and architecture.

Griffin, C.W. **Energy Conservation in Buildings: Techniques for Economic Design.** (Washington, D.C.: Construction Specifications Institute, 1974).

Overview of building energy conservation approaches and methods. Appendix contains a brief discussion on and examples for life-cycle costing in an energy context.

Griffith, J.W. "Applications of Engineering Economics to Integrated Lighting," Illuminating Engineering Magazine, Dec. 1962, Vol. LVII, No. 12, pp 785-790.

Illuminating Engineering Society. **IES Lighting Handbook,** 5th edition. (New York: Illuminating Engineering Society, 1972).

Illuminating Engineering Magazine, Daylighting Committee "Recommended Practice of Daylighting," August 1962, Vol. LVII, No. 8, pp. 517-557.

Libbey-Owens-Ford Co. **How to Predict Interior Daylight Illumination.** (Toledo, Ohio: Libbey-Owens-Ford Co., 1976).

A short "how to" booklet for use in starting to consider the feasibility of daylighting.

McGuinness, William J. and Benjamin Stein. **Mechanical and Electrical Equipment for Buildings,** 5th ed. (New York: John Wiley and Sons, 1971).

Textbook on fundamentals, concepts and calculation procedures for HVAC and lighting systems.

National Electrical Manufacturers Association and National Electric Contractors Association in cooperation with Federal Energy Administration. **Total Energy Management: A Practical Handbook in Energy Conservation and Management.** (Washington, D.C.: G.P.O., 1976).

Guidelines and checklists for improving energy efficiency in existing buildings at minimal expense.

Olgyay, Victor. Design With Climate: Bioclimatic Approach to Architectural Regionalism. (Princeton, N.J.: Princeton University Press, 1963).

Discuss regionalized architectural solutions based on scientific analysis and response to climatic conditions.

Public Technology, Inc. **Energy/Conservation Retrofit for Existing Public and Institutional Facilities.** (Washington, D.C.: 1977).

Directed toward public officials first considering energy conservation for existing buildings. Explains the processes involved except for the actual design and engineering methods.

Smith, Craig, ed. **Efficient Electricity Use.** (Elmsford, N.Y.: Pergamon Press, 1976).

Although focus is on electricity the document pulls together a massive amount of information on energy conserving approaches. Covers both building and urban applications.

State of California. **Energy Conservation Design Manual for New Non-Residential Buildings.** (Sacramento: Energy Resources Conservation and Development Commission, December, 1976).

An explanation and guide for California's non-residential energy standards which establish options for component performance requirements and building energy budgets.

Stein, Richard G. **Architecture and Energy: Conserving Energy through Rational Design.** (New York: Anchor Press, 1977).

A series of essays discussing the interralationships between issues which impact on the application and use of energy in building. Covers building design, environmental systems, and architectural principles for energy efficiency.

____________ and Carl Stein. **Research Design, Construction of Low Energy Utilization School.** (Report for the National Science Foundation, August, 1974).

Examines data base of energy consumption patterns for New York City schools and identifies the characteristics of schools which either use small or large amounts of energy compared with city wide average.

Stein, Richard G., Carl Stein and Paul F. Deibert. **Research, Design, Construction and Evaluation of a Low Energy Utilization School.** (Phase 2 report for the National Science Foundation, March, 1977).

Documents a series of reports on projects carried out to reduce demand in existing stock of New York City schools. Projects included operation manual; field visit recommendations; lighting and ventilation programs.

Trane Air Conditioning Manual. (La Cross, Wisconsin: Trane Co., 1977).

U.S. Air Force. **U.S. Air Force Manual 88-8.** Engineering Weather Data. (Washington, D.C.: G.P.O., 1967).

The essential source of weather data for performing energy analyses using temperature "bins."

U.S. Department of Health, Education and Welfare. **Total Energy Management for Hospitals.** (Rockville, Md.: H.E.W., 1977).

Guidelines and checklists for improving energy efficiency in non-critical areas of existing hospital facilities.

____________: **Total Energy Management for Nursing Homes.** (Rockville, Md.: H.E.W., 1977).

Guidelines and checklists for improving energy efficiency in existing small-scale nursing home facilities.

Van Straaten, J.F. **Thermal Performance of Buildings.** (Amsterdam: Elsevier, 1967).

Discusses a design procedure and non-mechanical physical features of buildings which affect indoor comfort. Based on research conducted by South African Council for Scientific and Industrial Research.

Watson, Donald. **Designing and Building a Solar House.** (Washington, D.C.: A.I.A., 1977). – Chapter 1 –